CLINICAL EXERCISE
PHYSIOLOGY
Laboratory Manual

PHYSIOLOGICAL ASSESSMENTS IN HEALTH, DISEASE AND SPORT PERFORMANCE

Second Edition

STEPHEN F. CROUSE

J. RICHARD COAST

WITH CONTRIBUTIONS FROM
GARY L. ODEN

Revised Printing

Kendall Hunt
publishing company

Cover art © 2011 Shutterstock, Inc.

Kendall Hunt
publishing company

www.kendallhunt.com
Send all inquiries to:
4050 Westmark Drive
Dubuque, IA 52004-1840

Copyright © 1995 by Wm. C. Brown Communications, Inc. A Times Mirror Company.

Copyright © 2011 by Kendall Hunt Publishing Company

ISBN 978-1-4652-1931-2

All rights reserved. No part of this publication may be reproduced,
stored in a retrieval system, or transmitted, in any form or by any means,
electronic, mechanical, photocopying, recording, or otherwise,
without the prior written permission of the copyright owner.

Printed in the United States of America
10 9 8 7 6 5 4 3 2

DEDICATION

To our wives, Jana and Holly, for their patience and love through our careers and the completion of this project.

CONTENTS

Preface	*vii*	
Chapter 1	Ergometry: The Measurement of Work and Power 1	
Chapter 2	Assessment of Speed, Anaerobic Power, and Anaerobic Capacity for Sport and Human Performance 15	
Chapter 3	Testing of Muscle Strength, Endurance and Flexibility 39	
Chapter 4	Pulmonary Function Testing 51	
Chapter 5	The Measurement of Oxygen Uptake and Energy Expenditure 65	
Chapter 6	Resting and Exercise Electrocardiography 77	
Chapter 7	Measuring Resting and Exercise Blood Pressure 107	
Chapter 8	Submaximal Exercise Testing for Estimating Aerobic Capacity 121	
Chapter 9	The Symptom-Limited Maximal Graded Exercise Test for Clinical and Sports Medicine Applications 141	
Chapter 10	The Measurement of Body Composition 183	
Chapter 11	Exercise Prescription for Aerobic Fitness in Healthy Populations 197	
Chapter 12	Exercise Prescription for Musculoskeletal Fitness and Health 221	
Appendix A	*Calibration of the Treadmill 233*	
Appendix B	*Calibration of the Cycle Ergometer 241*	
Appendix C	*Typical Calibration of Oxygen and Carbon Dioxide Analyzers 245*	
Appendix D	*Metabolic Calculations for Exercise Physiology 247*	

PREFACE

This revised laboratory manual is designed to be a practical teaching guide for training students and professionals in the skills to be applied to exercise testing and prescription for health and physical fitness. As such, it will be a valuable resource for exercise scientists, clinical exercise physiologists, sport physiologists, cardiac rehabilitation specialists, and physical educators at the undergraduate and beginning graduate levels. Indeed, the methods and procedures detailed will prove valuable to any professional who may need an understanding, practical background, or detailed procedural guide in evaluating the many dimensions of health, physical fitness, and human performance.

The major difference between this and other laboratory manuals in Exercise Physiology is the technique-based focus of this book. We firmly believe that learning the techniques is important, and therefore have concentrated on teaching the methodology. Where many laboratory manuals use designed experiments in which techniques can be applied, consequently providing less emphasis on *learning* the techniques, we have taken the position that learning the techniques in and of itself is a valuable learning objective. Students, in many cases, will have to apply the techniques to situations other than the experiments in which they were presented in many manuals. It is our belief that if the techniques are presented clearly so that the student or professional can master them, and the Laboratory Manual can also function as a review resource for those who need it, the student or professional will be better prepared to apply the techniques to real-life situations, including settings in clinical and sports exercise physiology and research. Where possible, we have included in the introductions details of how and when the techniques are appropriate and applied in clinical exercise testing, assessing human performance, sports medicine, and human exercise physiology situations.

ERGOMETRY: THE MEASUREMENT OF WORK AND POWER

CHAPTER 1

OUTCOME OBJECTIVES

Following successful completion of this exercise, the student should be able to:

1. Learn how work and power are measured, how the values for work and power are calculated, and how to convert from one measurement to the other.
2. Become familiar with the modes of exercise and practice the measurement of work and power when the work is accomplished by lifting the body weight, by such methods as stepping or walking on a treadmill.
3. Become acquainted with exercise testing and practice the measurement of work and power on devices that use resistance to create force, such as cycling machines.

INTRODUCTION

One of the backbones of exercise testing is the measurement of the actual work done and power output, as well as the energy produced by the human body. Measurements of strength and power have been made for centuries (Regnier, 1798). The knowledge of how much energy a subject can produce is often of little meaning without the accompanying knowledge of how much work that subject can do with that energy production. In one of the earliest reports of an accurate measurement of efficiency (Efficiency = work done/energy consumed to perform that work), Scoresby and Joule (1846) measured the work done and the food consumed by a horse in an average workday. They found that approximately 20% of the energy in the food consumed was accounted for in the work performed. This efficiency estimate is comparable to those measured today with the most sophisticated techniques. In fact, the Systeme Internationale unit for energy is named after Joule. These measurements are not new, as evidenced by the examples above, but are still of great importance.

Work is simply the application of force over a distance, or:

$$\text{Work} = \text{Force} \times \text{Distance}$$

The work done divided by the time taken to perform that work is known as power. It can be expressed as below:

$$\text{Power} = \text{Work}/\text{Time}$$

Or, if you are using velocity, as in walking on a treadmill, since velocity is expressed by the following equation:

$$\text{Velocity} = \text{Distance/Time}$$

you can combine the two equations as follows:

$$\text{Power} = (\text{Force} \times \text{Distance})/\text{Time} \text{ or Power} = \text{Force} \times \text{Velocity}$$

Common units of work and power are given below as well as some of the conversions between the various units.

TABLE 1.1— Common units of work or energy, and power output, and the conversions between them

\multicolumn{3}{c}{WORK AND ENERGY}		
Kilopond-meter (kp.m)	=	7.233 ft.lbs 9.81 joules 0.00234 kcal
Kilogram-meter (kg.m)	=	1 kp.m
joule (j)	=	1 Newton-meter (N.m) 0.101 kp.m 0.737 ft.lbs 0.000238 kcal 0.001 kilojoules (kJ or J)
Kilocalorie (kcal)	=	426.8 kp.m 3087.4 ft.lbs 4186 joules 4.186 J
\multicolumn{3}{c}{POWER}		
Kilopond-meter per minute (kp.m.min^{-1})	=	0.163 watts 0.00234 kcal/min 9.81 joules/min 7.233 ft.lbs/min 0.00022 horsepower 1 kpm/min
Watt (W)	=	1 N.m/sec 6.12 kp.m/min 44.24 ft.lbs/min 0.01434 kcal/min 0.00134 horsepower
Horsepower (hp)	=	745.7 W 4564 kp.m/min 10.69 kcal/min 33,000 ft.lbs/min 44.743 joules/min

In the exercise physiology laboratory, when we measure the work done by a subject, we can measure only the external work. There is a certain amount of work done by the subject that we do not measure routinely. An example of this is in running. With each step the runner's center of gravity moves up when the runner lifts off the ground. The movement of the center of gravity requires work, but without the aid of highly quantitative cameras and image analysis systems, that work cannot be measured. The same is true in cycling, where much of the work done is expended to overcome air resistance, which cannot be measured without special equipment. Therefore, in the laboratory we use equipment in which the measurable work is maximized and the unmeasured work is decreased as much as possible. Of the ergometers used in laboratories on a regular basis, two use the body weight (gravity) to exert the force and measure the distance climbed as the distance component. Examples of these are the treadmill and the stepping ergometer. Another uses resistance to create the force and measures the distance by how far a spot on the outside of a flywheel travels. This combination of force and distance constitutes the work done. This is exemplified by the cycle ergometer.

Many of the activities that we do for recreation or for fitness are mimicked to a certain degree with the ergometers. Current cycle and rowing ergometers simulate bicycling and rowing quite well and thus provide a very good sport-specific testing mode. Treadmills are of all types, from those that will allow us to do little more than walk or jog slowly at a low incline to those capable of testing elite athletes and animals such as horses and antelopes. Stepping has generally been thought of as a fitness testing mode for soldiers, without much real-life application. The step has come into much more widespread use in aerobic exercise classes, though, and is now considered a more activity-specific ergometry method than might have been thought previously.

Why would we need to measure work and power? The uses for ergometry are many. In a clinical situation, it is common to perform submaximal exercise tests on either a cycle ergometer or a treadmill. The investigator generally measures the power, heart rate, and oxygen uptake at several successive submaximal loads, and plots these on a graph to get an estimate of the maximal capacity. From these, exercise prescriptions are given. These tests are described in other chapters. To estimate the correct maximal workload and to calculate the correct exercise prescription, these tests depend on several measurements of work and power. In the workplace, industrial engineers spend a great

FIGURE 1.1 — Photographs of three common exercise testing modalities, the cycle ergometer, the treadmill, and the step.

© Julián Rovagnati, 2011. Used under license from Shutterstock, Inc.

© Lisa F. Young, 2011. Used under license from Shutterstock, Inc.

© Diego Cerva, 2011. Used under license from Shutterstock, Inc.

deal of time measuring the work that a person can do in a workday, whether that work involves laying bricks, assembling automobiles, or loading packages onto a truck. These measurements are done both to protect the worker from having to perform so much work that he or she will be injured on the job and to help management maximize the productivity of the worker. Accurate measurements of work and power also help researchers who work with athletes to improve their performance capacities.

In this exercise you will be introduced to the measurement of work and power on three common pieces of equipment: the step, the treadmill, and the cycle ergometer. Because these are the most common ergometric apparati, they will be the ones on which we concentrate. Many other ergometers are also used in very sport specific situations, and most operate in a similar manner to one of the ones we will examine here. After you have completed these exercises, you should be able to understand the use of many other types of ergometers as well.

BENCH STEPPING

Equipment Needed

Sturdy bench or step

Stopwatch or clock timer

Metronome

Meter stick

Scale for weighing subject

TECHNIQUES AND PROCEDURES

The step test is one of the simplest tests that is routinely used in the laboratory, and as previously mentioned, it is now a relatively widely used test. When stepping, the work done is the product of the weight lifted and the distance traveled in the upward direction. Therefore, the work performed during stepping is dependent on body weight. In these tests, work is measured only against gravity. Therefore, the "negative" work performed when stepping down is not measured. Ignoring the "negative" work is all right, though, because it requires very little metabolic energy to perform and does not contribute very much error (LaStayo et al., 1999). To perform a measurement of work and power during stepping, use the following sequence of instructions:

1. Set up a step or bench. A number of the standard tests use a bench of 12 inches in height (ACSM, 2010, p. 80). Although bench height may alter responses to stepping, this standard is a good starting point. Regardless of what step height you use, the bench should be sturdy enough to easily support the weight of the subjects who will be using it and should have surfaces that will prevent the subject from slipping on the bench or the bench from slipping on the floor. Record the exact height of the bench (in meters) on your worksheet.

2. Weigh the subject to the nearest 0.5 kg. The weight that you measure should include anything the subject will wear during the test since clothing or other apparatus such as backpacks contribute to the load being lifted. This weight should also be recorded on your worksheet. While scales measure in pounds or kilograms, the appropriate unit of force is the kilopond (kp), which is the force exerted by a 1 kg mass.

3. Set the metronome at the desired cadence. The metronome rate should be four times the rate at which you want the subject to complete one cycle of stepping. In the YMCA procedure, the subject steps up and down 24 times per minute, so the metronome should be at 96 beats per minute (Golding, Myers, and Sinning, 1989). Other tests may have the subject working at any cadence between 10 and 30 steps per minute. Cadence should also be recorded on your worksheet.
4. Have the subject step onto the step and back down, one movement per beat ("up-up-down-down"). To help avoid muscle soreness, it is often preferable to have the subject switch legs approximately halfway through the test.
5. Count the number of complete steps that the subject takes and measure the time it takes to perform the test. A good time to use is three minutes at any combination of step height and rate. Record these values on your worksheet in the appropriate blanks.
6. Calculate the work and power using the following formulae and on the worksheet:

$$\text{Work (kp.m)} = \text{Step height (m)} \times \text{number of steps performed} \times \text{body weight (kg)}$$

1 in = 0.0254 m

$$\text{Power (kp.m/min)} = \text{Work performed (kp.m)/time taken to perform the work (min)}$$

or

$$\text{Power (kp.m/min)} = \text{Number of steps/minute} \times \text{step height (m)} \times \text{body weight (kg)}$$

Repeat the measurements with several different bench heights or rates of stepping. Measure work and power at four different combinations of step height and rate. Try measuring pulse rate at the wrist for 10 seconds just prior to stopping each work bout. Graph the heart rate against the power output.

TREADMILL WALKING

Equipment Needed

Treadmill

Timer

Scale for weighing subject

TECHNIQUES AND PROCEDURES

As with stepping, work on a treadmill is performed against gravity only if the treadmill is inclined. Work in the horizontal direction is very difficult to measure. Therefore, to be able to measure the work done, you will need to know the distance ascended, which depends on both the distance that the treadmill belt has moved and the grade, as well as the subject's body weight, which is the force component of the work formula. Calibration of treadmill speed and grade is critical to knowing the work performed. This calibration is described in Appendix A.

The treadmill elevation is generally reported as the angle of incline or the percent grade. If you picture the relationship that the treadmill makes with the floor (Figure 1.1), it can be viewed as a

FIGURE 1.2 — Diagram of treadmill showing the relationship of angle, percent grade, sine, and tangent. Tangent θ = Grade = Rise/Run. % Grade = Grade × 100. Sine θ = Rise/Hypotenuse.

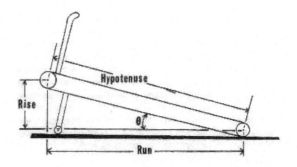

right triangle where the percent grade is the rise over the run, or the tangent of the angle θ. However, the distance the belt travels and the horizontal distance (the run) are not the same. Therefore, the distance ascended is not simply the product of the speed and the percent grade in most instances. To compute the distance ascended, you will need to use the distance the belt travels multiplied by the sine of the angle θ (rise over belt length). When the treadmill grade is less than 10%, the sine and the tangent are nearly identical, so either fraction can be used, but as the angle increases, the two fractions become more and more different, and the sine of the angle is the correct term to use to compute the distance ascended.

Therefore, work on the treadmill is calculated by the following formula:

$$\text{Work (kp.m)} = \text{Distance traveled (m)} \times \sin\theta \times \text{body weight (kg)}$$

and

$$\text{Power (kp.m/min)} = \text{Work (kp.m)} \div \text{time (min)}$$

or

$$\text{Power (kp.m/min)} = \text{Speed (m/min)} \times \sin\theta \times \text{body weight (kg)}$$

The conversion between percent grade (tangent θ), angle θ, and sine θ can be obtained from table 1.2.

To measure work and power on the treadmill, follow the procedures outlined below:

1. Weigh the subject to the nearest 0.5 kg. As in the tests using the step, find the weight of the subject while he or she is wearing all clothing and equipment that will be worn during the test, since this will have to be lifted, as will the body weight.
2. Turn the treadmill on and adjust it to a slow speed (e.g., 2 mph). Be sure that the subject is not standing on the treadmill belt when it is started.
3. Have the subject straddle the treadmill belt and paw the belt with one foot while still holding on to the handrails. This helps the subject get comfortable with the speed. When comfortable, have the subject begin walking on the treadmill and eventually release the rails.
4. Gradually increase the treadmill to the desired speed and grade, and record the speed and percent grade in the appropriate spaces on the worksheet. A typical speed and grade at which to start a test is 3 mph and 2.5% grade.
5. Measure the time that the subject walks at this speed and grade, and record it on your worksheet.
6. Calculate the work done and the power output using the formulae provided earlier.

TABLE 1.2 — Conversions between treadmill grade, angle of incline, and sine

% Grade	(Tangent θ)	Angle θ (degrees)	Sine θ
1	(.010)	0.573	.010
2	(.020)	1.146	.020
2.5	(.025)	1.432	.025
3	(.030)	1.718	.030
4	(.040)	2.291	.040
5	(.050)	2.862	.050
6	(.060)	3.434	.060
7	(.070)	4.004	.070
7.5	(.075)	4.289	.075
8	(.080)	4.574	.080
9	(.090)	5.143	.090
10	(.100)	5.711	.100
12	(.120)	6.843	.119
12.5	(.125)	7.125	.124
14	(.140)	7.970	.139 [ang out]
16	(.160)	9.090	.158
17.5	(.175)	9.926	.172
18	(.180)	10.204	.177
20	(.200)	11.310	.196
22.5	(.225)	12.680	.220
25	(.250)	14.036	.243

Try this at several different speeds and grades while measuring heart rate after 2 minutes at each level. For this exercise, it is better to use relatively low speeds and increase the grade rather than just increase the speed. That maximizes the measureable work that is done. Why? Graph the heart rate against the power output.

CYCLE ERGOMETER

Equipment Needed

Cycle ergometer capable of accurate resistance setting

Timer

Metronome

TECHNIQUES AND PROCEDURES

As opposed to the treadmill or the step, work on the cycle ergometer does not depend on the body weight of the subject, as the subject is supported by the ergometer itself. Instead, work depends on the resistance applied to the ergometer flywheel, and the distance traveled by a point on the perimeter of the flywheel.

This often makes the cycle ergometer the instrument of choice for testing subjects who are severely obese or who have muscle weakness or problems with balance, subjects who would have a difficult time supporting themselves.

To measure the work performed on a cycle ergometer you will need to know a number of variables. These include: (1) the resistance applied to the flywheel; (2) the number of pedal revolutions; and (3) the distance the flywheel rotates with each pedal revolution. On many popular cycle ergometers, the resistance is applied by tightening or loosening a friction belt around the flywheel. This is the most common calibratable method, although a number of ergometers also use electronic braking to provide resistance. Calibration of the friction belt-braked cycle ergometer is described in Appendix B. The number of pedal revolutions is easily counted; or you can have the subject pedal at a standard cadence and multiply the pedal rate (revolutions per minute) by the time pedaled to obtain the total number of revolutions. The distance the flywheel travels with each pedal revolution is a constant. For Monark® ergometers this constant is 6 meters per pedal revolution. Therefore, the work performed on the cycle ergometer can be calculated using the formula:

$$\text{Work (kp.m)} = \text{Number of pedal revolutions} \times 6 \text{ m/rev} \times \text{resistance (kp)}$$

and the power is expressed as:

$$\text{Power (kp.m/min)} = \text{Work (kp.m)} \div \text{time (min)}$$

or

$$\text{Power (kp.m/min)} = \text{Pedal rate (rev/min)} \times 6 \text{ m/rev} \times \text{Resistance (kp)}$$

To evaluate work and power output on the cycle ergometer, perform the following procedure:

1. Adjust the seat height to fit the subject. It should be adjusted to a height where the subject's heel can be placed comfortably on the pedal when the pedal is at the bottom of the cycle.
2. Set the metronome to the proper rate. Most subjects find it easier to pedal in time with a metronome that is set at 2 times the specified pedal rate. Typical pedaling rates for laboratory settings are 50 or 60 RPM (100 or 120 metronome beats per minute) (ACSM, 2010, p. 76) while for trained cyclists you might want to use a higher cadence such as 80 RPM (Coast and Welch, 1985).
3. Have the subject start pedaling at the set cadence and set the resistance to the desired load using the resistance knob. A typical starting point for many tests is 60 RPM and 1 kp resistance. The resistance can be read off of the scale on the side of the friction braked ergometer or from a readout on an electrically braked ergometer. Record both the pedal rate and the resistance setting in the appropriate blanks on your worksheet.
4. Either count the number of pedal revolutions that the subject performs or record the time that the subject pedals at the specified cadence. Record these values on your worksheet.
5. Using the equations above and on your worksheet, calculate the work performed by the subject on the cycle ergometer.

Try having the subject pedal the ergometer during 2–3 loads for two minutes at each load. At the end of the two minute period, record the subject's heart rate. Plot the heart rate against the power output for each load.

QUESTIONS AND ACTIVITIES

1. Calculate work and power for each of your subjects on each modality at each workload.
2. Graph HR versus power output for each subject on the step, cycle and treadmill test, with power output on the x axis.
3. Is there a relationship between power output and heart rate for stepping? ... on the treadmill? ... on the cycle ergometer?
4. Is the relationship between power and heart rate quantitatively similar for the three exercise modes? Can you suggest a reason why or why not?
5. Would you expect the heart rate-to-power output ratio to the same if you achieved the power using a slow speed and large change in grade on the treadmill compared to a low grade and large change in speed? Why or why not?
6. On the step, negative work is done that is not measured. How would this affect the power-to-heart rate relationship?

REFERENCES

American College of Sports Medicine. *ACSM's Guidelines for Exercise Testing and Prescription,* 8th ed., Philadelphia: Lippincott, Williams and Wilkins, 2010.

Coast, J. R. and H. G. Welch. "Linear Increase in Optimal Pedal Rate with Increased Power Output in Cycle Ergometry." *Eur. J. Appl. Physiol.* 53 (1985): 339–42.

Golding, L. A., C. R. Myers, and W. E. Sinning (eds.) *Y's Way to Physical Fitness,* 3rd ed., Champaign, IL: Human Kinetics Publishers, 1989.

LaStayo, P. C., T. E. Reich, M. Urquhart, H. Hoppeler, and S. L. Lindstedt. "Chronic Eccentric Exercise: Improvements in Muscle Strength Can Occur with Little Demand for Oxygen." *Am. J. Physiol.* 276 (1999): R611–R615.

Regnier, C. "Description and Use of the Dynamometer, or Instrument for Ascertaining the Relative Strength of Men and Animals." *Phil. Mag.* 1 (1798): 399–404.

Scoresby, W. and J. P. Joule. "Experiments and Observations on the Mechanical Powers of Electro-magnetism, Steam, and Horses." *Phil. Mag.* 28 (1846): 448–55.

30.5 cm

STEPPING WORKSHEET

18, 20, 24 [Allegra]

Audrey At Rest 84

Subject weight __67.5__ kg (lbs ÷ 2.205)   52.6 kg / 515 N

Candace At rest 60

Subject #2 weight __48.0__ kg

Step height   Trial 1 __0.30__ m (in × 0.0254)
              Trial 2 __0.30__ m
              Trial 3 __0.30__ m

Step Height 1. 0.30
             2. 0.30
             3. 0.30

Step rate   Trial 1 ~~18~~ 36 steps/min
            Trial 2 ~~20~~ 40 steps/min
            Trial 3 ~~24~~ 48 steps/min

Step Rate 1. ~~18~~ 36
           2. ~~20~~ 40
           3. ~~24~~ 48

Steps taken __124__ steps (or step rate × minutes)

Steps Taken 124

Work done (kp.m) = Step height (m) × Number of steps × body weight (kg)
   = __.30__ m × __36__ × ~~67.5~~ kg   Do all Work Done:
                              515 N   36, 40, 48
   = ~~724~~ kp.m   5,562 N·m

Power output (kp.m/min) = Work done (kp.m) ÷ time (min)   Power Output:
   = ~~724~~ 5562 N·m kp.m ÷ __120__ min s
   = ~~301.5~~ kp.m/min   46.35 W

100 bpm
120 bpm
132 bpm

or you can use the equation:

Power output (kp.m/min) = Step rate (steps/min) × step height (m) × body weight (kg)
   = _____ steps/min × _____ m × _____ kg   Power Output:
   = _____ kp.m/min

  ★ Repeat calculations for Trials 2 and 3

Heart rate (bts/min)   Trial 1 ~~IIII~~ 100
                       Trial 2 ~~IIII~~ 120
                       Trial 3 ~~IIII~~ 132

Heart Rate 1. ~~IIII~~ 116
            2. 120
            3. ~~IIII~~ 136

Power output (kp.m/min)   Trial 1 __46.35 W__
                          Trial 2 __51.5 W__
                          Trial 3 __61.8 W__

Power Output 1.
             2.
             3.

11

TREADMILL WORKSHEET

Faith
65 kgs 637N

Subject weight __86.6__ kg (lbs ÷ 2.205) HR=80

Treadmill speed (m/min)	Trial 1 __80.5__ m/min	Treadmill grade	Trial 1 __5__ %
= MPH × 26.822)	Trial 2 __80.5__ m/min		Trial 2 __10__ %
	Trial 3 __80.5__ m/min		Trial 3 __15__ %
Angle θ	Trial 1 __2.862__ deg	Sine θ	Trial 1 __.050__
	Trial 2 __5.711__ deg		Trial 2 __.100__
	Trial 3 __8.53__ deg		Trial 3 __.149__
Walking time	Trial 1 __2__ min	Walking distance	Trial 1 __161__ m
	Trial 2 __2__ min	(Speed × Time)	Trial 2 __161__ m
	Trial 3 __2__ min		Trial 3 __161__ m

Work done (kp.m) = Walking distance (m) × sine θ × body weight (kg)

= __161__ m × __0.050__ × ~~86.6~~ __637 N__ kg   Do all = 697.1 kp·m not 629.1

= _____ kp.m   5,127.85 N·m   other ones are right

Power output (kp.m/min) = Work done (kp.m) ÷ time (min)

= __5,127.85__ ~~kp.m~~ N·m ÷ __120__ min ✓

= __42.73__ ~~kp.m/min~~ W

or the following equation can be used:

Power output (kp.m/min) = Speed (m/min) × sine θ × body weight (kg)

= _____ m/min × _____ × _____ kg

= _____ kp.m/min

88 bpm
80 bpm ?
116 bpm

Repeat calculations for Trials 2 and 3

Heart rate (bt/min)	Trial 1 ~~108~~ 88	HR	1. 104
	Trial 2 ~~112~~ 80 ?		2. 120
	Trial 3 ~~120~~ 116		3. 136
Power output (kp.m/min)	Trial 1 __42.73 W__	DO	1.
	Trial 2 __85.46 W__		2.
	Trial 3 __127.34 W__		3.

Subject Weight __85.6 kg__   HR=64

Treadmill Speed 1. 80.5
2. 80.5
3. 80.5

Angle θ 1. 2.862
2. 5.711
3. 8.53

Walking Time 1.
2.
3.

Treadmill Grade 1. 5 %
2. 10 %
3. 15 %

Sine θ 1. .050
2. .100
3. .149

Walking Distance 1.
2.
3.

CYCLE ERGOMETER WORKSHEET

Audrey 99.5 kg 976

Pedal rate Trial 1 ~~100~~ 50 rev/min 60   Resistance   Trial 1 __1__ kp 0 9.81 (9.81 N
              Trial 2 ~~100~~ 50 rev/min 60   Trial 2 __2__ kp   19.62 N
              Trial 3 ~~100~~ 50 rev/min 60   Trial 3 __3__ kp   29.43 N

Work time   Trial 1 __2__ min   Pedal revolutions   Trial 1 __100__ revolutions 120
              Trial 2 __2__ min   (rev/min × time)   Trial 2 __100__ revolutions 120
              Trial 3 __2__ min   Trial 3 __100__ revolutions 120

Work done (kp.m) = Number of pedal revolutions × 6 m/rev × resistance (kp)
    = __120__ revolutions × 6 m/rev × __9.81__ kp   7,063.2

Power output (kp.m/min) = Work done (kp.m) ÷ work time (min)
    = __7,063.2__ kp.m ÷ __120__ ~~min~~ s
    = __58.86__ kp.m/min

or you can use this equation:

Power output (kp.m/min) = pedal rate (rev/min) × 6 m/rev × resistance (kp)
    = _____ rev/min × 6 m/rev × _____ kp

✗ Repeat calculations for Trials 2 and 3

Heart rate (bt/min)   Trial 1 __100__
                          Trial 2 __1¢2__
                          Trial 3 __120__

Power output (kp.m/min)   Trial 1 __58.86 W__
                          Trial 2 __117.72 W__
                          Trial 3 __176.58 W__

100 bpm
112 bpm
120 bpm

Kate

Pedal Rate 1. ~~100~~ 50
            2. ~~100~~ 50
            3. ~~100~~ 50

Resistance 1. 1
            2. 2
            3. 3

Work Time   1. 2
            2. 2
            3. 2

Pedal Revolutions 1. 100
                   2. 100
                   3. 100

Work done:

Power Output:

Power Output:

Heart Rate 1. ~~100~~ 100
            2. 104
            3. 116

Power Output 1.
              2.
              3.

CHAPTER 2

ASSESSMENT OF SPEED, ANAEROBIC POWER, AND ANAEROBIC CAPACITY FOR SPORT AND HUMAN PERFORMANCE

OUTCOME OBJECTIVES

Following successful completion of this exercise, the student should be able to:

1. Understand and describe the type of exercise that best exemplifies the predominant use of anaerobic phosphagen and glycolytic energy systems.
2. Perform and interpret tests that evaluate speed, short-term anaerobic and moderate-term anaerobic capacities.
3. Choose physiologic tests of speed, agility, and anaerobic power and capacity that are specific to sport or occupational tasks.
4. Relate speed and anaerobic test results to training requirements for sport and human performance.

INTRODUCTION

Health-related physical fitness has the greatest potential for widespread benefit to the population, and is typically the goal of worldwide public health policy. The "Exercise Is Medicine" initiative, a product of the combined efforts of the American College of Sports Medicine and the American Medical Association, is witness to the broad-based efforts to integrate physical activity and exercise into the medical model for disease prevention and medical treatment (American College of Sports Medicine, 2011). Since cardiovascular endurance and healthy body composition are emphasized in this model, tests to assess cardiovascular function, body composition, respiratory status, and muscle fitness predominate in the physical fitness domain. It is also true that many of the traditional physical fitness and human exercise physiology measurements have been designed to evaluate exercise in the steady state, which is by definition aerobic. Therefore, the majority of physical fitness/human performance tests, and consequently those detailed in the chapters in this laboratory manual, involve testing or exercise prescription procedures for the aspects of health-related physical fitness.

In contrast to the predominance of aerobic (endurance) activities recommended for health, the sports that dominate the world's attention and have the greatest participation rates among youth and young adults generally demand speed, agility, and muscle power for optimal performance. Examples of these types of sports include basketball, soccer, American football, baseball, many track and field events, and downhill skiing. Furthermore, professionals in certain occupations are benefited in their work performance by the physical attributes of speed, agility, and the ability to exert short term, near maximal efforts. Military, fire-protection, and police professionals, for example, require the ability to perform very strenuous tasks over short durations. All sports and activities in which speed and power are essential to success demand a large contribution from anaerobic energy systems. Measuring the capacity of these anaerobic systems and devising adequate training protocols to improve them are major goals of athletes, coaches, and sports physiologists. Several tests of speed and anaerobic exercise fitness will be described in this laboratory manual chapter.

In order to measure anaerobic fitness and capacity, and the derivative speed and agility, it is important to understand what does and does not constitute anaerobic exercise. The term *anaerobic* refers to any of the metabolic pathways that use fuels, such as glucose, to generate usable energy without oxygen as the final electron (hydrogen) accepter, which is the case during "aerobic" metabolism. In humans at rest and during low-intensity aerobic exercise, fats and to a lesser extent carbohydrates are the predominant fuels used to produce energy in the form of adenosine triphosphate (ATP), with water and carbon dioxide as the final products of the "aerobic" Krebs cycle and electron-transport system. The rate of energy flux through this cellular machinery, however, is unable to supply the rapid and elevated ATP needs at the beginning and during relatively short-term (up to about 5 min) high-intensity exercise. In this case, anaerobic energy metabolism predominates to quickly produce the ATP required, although aerobic metabolism is not "shut off" completely even during heavy physical work, and recovery from high-intensity work is again predominately aerobic.

One of the primary sources of energy for very explosive, short-term (up to 30 sec), maximal exertion is that derived from the phosphagen system, comprised of ATP and creatine phosphate (CP). These two molecules are referred to as phosphagens because both contain high-energy phosphate bonds. Within limits, the more phosphagens stored in muscle, the better a person is able to perform short-term, high-intensity exercise, such as sprinting. Some nutrient supplements, such as creatine, are aimed at enhancing the intramuscular storage of phosphagens, and have been shown have ergogenic benefits under some conditions (Rawson and Volek, 2003). The human fast twitch (type IIa and IIx) muscle fibers contain relatively higher amounts of the phosphagens and the enzymes with which to break them down. The phosphagen system, although to some extent genetically determined, can be improved with training (Powers and Howley, 2009).

The other primarily anaerobic energy system is anaerobic glycolysis. Glycolysis occurs under both aerobic and anaerobic conditions, but when energy demand exceeds oxidative metabolism of the muscle, stored glycogen can be quickly converted to glucose, which is subsequently shuttled through the glycolytic pathway resulting in the rapid production of ATP without oxygen. Anaerobic glycolysis is capable of producing sufficient supplies of ATP very quickly to fuel high-intensity exercise demands, but only for a relatively short period of time. The by-product of this anaerobic metabolism of glucose is an increase of lactic acid over that normally present in the cell at rest or during light exercise. In sufficient quantities, lactic acid will cause fatigue, and an obligate reduction in exercise intensity as muscle contractile ability becomes limited. This metabolic pathway is predominant in maximal-effort exercise bouts lasting between 0.5 and 5 minutes, and neither fats nor proteins can be used as fuels (Powers and Howley, 2009). The glycolytic machinery is especially dominant in the fast twitch muscle fibers, and since muscle fiber type distribution is largely genetically determined, genetics appears to play a role in the selection of people with a high glycolytic capacity. However, specific physical training practices can increase anaerobic power through increasing muscle size, altering fiber types, substrate storage, anaerobic enzyme activity, and increased tolerance and removal of metabolic by-products, particularly lactic acid.

Several tests of running, jumping, stepping, and cycling will be described in this laboratory chapter that are useful in assessing speed and the short-term and moderate-term anaerobic capabilities of athletes or nonathletes. More emphasis will be placed on the measurement techniques themselves than on using the values obtained to categorize subjects into high, average, or low speed or anaerobic fitness levels. Indeed, such categories are of limited value unless used to compare individuals within a specific occupation or sport, since many of the tests are applied to athletes in sports or persons in professions, which vary widely in power and speed requirements for success.

PROCEDURES FOR SPEED, AGILITY, AND SHORT-TERM ANAEROBIC POWER

Speed, agility, and short-term anaerobic power (or capacity) are related to success in many sports, and also contribute to job-related performance in certain military and public-protection occupations. Training these components of anaerobic fitness are goals of many coaches, athletes, and training officers in the military. The tests demand maximal effort and are of short duration, lasting less than 20 seconds, and therefore maximally challenge the phosphagen system, with little contribution from anaerobic glycolysis. Undoubtedly neuromuscular coordination plays a role in the performance of these tasks, but this factor is very difficult to measure independently. However, training protocols often focus on neuromuscular training as well as physiologic and metabolic improvement.

ANAEROBIC SPRINT TESTS

Sprint tests at various distances are some of the most popular anaerobic power tests given in field situations and applied to sports performance. These tests are relatively easy to administer, are reproducible, and require minimal equipment, and they are directly related to successful performance in many sports. It is a widely held opinion born out in practice that speed (i.e., maximal velocity of running) is an asset to athletic performance in most sports. Hence, the importance of measuring and improving speed. Sprints of 40 yards are commonly used in the assessment of speed for many athletes, most notably football athletes. Other distances may be used in other sports where different sprint distances make logical sense, but the measurement procedures remain essentially the same. For example, 30- or 60-yard sprint times are often assessed in baseball players, 20-yard times in women softball players, and the 60-meter dash time reflects the ability of pure sprinters in track and field. Since a 40-yard sprint usually takes less than about 8 seconds to perform, even for younger subjects, it is considered almost exclusively a test of the phosphagen system.

As with most testing activities in this laboratory manual, the procedures are most efficiently carried out by a team of at least three individuals to complete the procedures, record the necessary data, serve as the subject for the tests, and complete the summary activities. Be sure all timers are familiar with the stopwatch operation. There is a learning curve in the reaction time required to start and stop the stopwatch, so some practice time before the real measures are taken is highly recommended. Note that the index finger should always be used on the start-stop switch on the stopwatch.

Equipment Needed

Soft-surface running area at least 20 yards longer than the test distance

Stopwatch for manual timing (or electronic timing instrumentation)

Measuring tape

Whistle

Marking tape and cones

Test Preparation

1. Prior to the arrival of the subject, make sure that the sprint course is accurately measured out and clearly marked. If electronic timing equipment will be used, be sure it is set up and checked for operational accuracy. For this laboratory, 40 yards will be the test distance. Often, 10-yard increments are marked along the course. If testing is to be performed on a football field or an indoor track, distances may be premarked, but these should be measured for accuracy.
    a. Carefully walk the proposed course to be sure it is completely free of holes, unevenness in the surface, sprinkler heads, small drains, etc., anything which could pose a tripping hazard and cause an injury. Move the course as necessary to avoid any dangers to the runners.
    b. Use a measuring tape to accurately measure the required distance in yards or meters. Mark the starting line, each 10-yard increment, and the finish line using a line of tape on the running surface and visible cones on each side. Be sure to allow at least 20 yards at the end of the course as a safe slow-down area.
2. Communicate the procedures and expectations to the subject prior to arrival.
    a. Clothing should include loose-fitting exercise shorts and tops. Shoes should be lightweight running shoes suitable for the surface on which the test will be run. Athletes usually wear the specific shoes for their sport when at all possible.
    b. No food should be consumed for at least 2 hours before reporting for testing, and drinks should be restricted to water only.

Test Procedure

1. Upon arrival, the technician should measure the subject's height to the nearest 0.5 cm and weight to the nearest 0.25 kg, including any clothing that will be worn during the test. Record these data on the Speed, Agility, and Short-Term Anaerobic Power Data sheet.
2. Just prior to testing, lead the subject through a series of stretching and warm-up activities, including jogging/running, gradually increasing in intensity and carefully designed to prepare the subject for the all-out sprint task.
    a. As for all testing procedures, which may be repeated serially over the course of a training season or year, the warm-up should be standardized between repeat measures, since variation in warm-up activities could affect the outcome of the test administered weeks or months apart. In addition, sufficient warm-up is a critical element of injury prevention.
3. Station one laboratory team member to monitor the starting line and another at the finish line to time the sprint, OR additional timers as follows.
    a. In many sports the "quickness" (acceleration) in which an athlete reaches maximal velocity is important. This can be measured with some degree of accuracy by timing 10-yard increments throughout the sprint, or at least the first 20 yards. This will require stationing additional timers at each 10-yard increment as desired.

4. After sufficient warm-up, the laboratory team member responsible for supervising the start of the sprint should escort the subject to the starting line, explain the starting procedure, encourage them to give an all-out speed effort sprinting *through* the finish line, and assist them in assuming the correct starting position. In this laboratory, a stationary start position will be used.
   a. It is most common for the subject to assume a three-point football stance or four-point sprinter stance with hand(s) just behind the starting line, weight on the balls of his or her feet spread comfortably about shoulder width apart, weight shifted slightly forward onto the hand(s).
   b. A two-point standing start with one foot just behind the starting line may be elected if specific to the sport (e.g., baseball), but this will typically result in comparatively slower sprint times.
   c. A running start may be permitted, depending on the sport application, but timing of the test is much more difficult, unless electronic timing devices are available.
5. Once the subject runner is at the starting line, the start sequence can be initiated in one of two ways.
   a. The most common approach in sport testing is for the starter to verbalize the following commands:
      i. The "On your mark…" command calls the runner to assume a comfortable, relaxed starting position at the starting line, usually on the knees with the hand(s) just behind the starting line.
      ii. The "Get set" command, calls the runner to assume and hold the starting position ready to initiate the sprint. At the same time the starter should raise an arm above the shoulder to alert the timers to be prepared to start timing.
      iii. After a brief hold in the ready position (no "rolling" starts permitted) the subject starts the sprint at the time of the runner's own choosing.
      iv. The timers are instructed to carefully observe the subject and **start** timing on the first observed subject movement after the "Get set" command is visually confirmed.
      v. The starter should call a "False start" by loudly sounding a whistle immediately if the runner fails to execute the start as prescribed, and the runner will be allowed to restart the sprint.
   b. A second approach is to use a typical track starting series of "On your mark, Get Set, GO!" The team member in charge of the start should loudly verbalize these commands, raise one arm above the shoulder at the "Get Set" command, and swing the arm forcefully downward on the "GO" command.
      i. The timers are instructed to be ready to start the stopwatch when the starter's arm is raised, and **start** timing at first movement of the starter's arm drop.
6. The timers start the stopwatch(s) on the first movement of the subject or at the first movement of the starter's arm as described above. The stopwatch is stopped by the timer when any part of the runner's body crosses the plane of the timing line if split times are done in 10-yard increments, and always at the finish line.
7. Record the sprint distance, split times (if used) and the finish time on the Speed, Agility, and Short-Term Anaerobic Power Data sheet. Because 40 yards is the sprint distance for this laboratory, there will be four split times, 0–10, 10–20, 20–30, 30–40 yards.
   a. Calculate the split velocities by dividing the split duration in seconds into the split distance, 10 yards. The split duration is calculated by subtracting the stopwatch time recorded by the timer stationed at the end of the previous 10-yard split from the stopwatch time recorded for the current split. For example, for the split duration for the 20–30 yard split, the stopwatch time recorded by the timer at 20 yards would be subtracted from the stopwatch time recorded by the timer at 30 yards. The unit of measure will be $yds \cdot sec^{-1}$ for each split. This can be converted to $ft \cdot sec^{-1}$ by multiplying by a factor of 3 if desired.

    b. The velocities for each split can then be plotted to obtain a velocity profile over the entire sprint distance and analyzed for the maximal velocity the runner attained, the split in which it was reached, and how much the runner decelerated afterward.

8. Generally two or three efforts are given. A third effort may not be needed if the times for the first two tests are within 0.2 sec of each other. Be sure to allow time for full recovery between tests. Since this test stresses the phosphagen system, full recovery requires about 3–5 minutes (Powers and Howley, 2009).
9. After the final test, direct the subject in cool-down routines, including slow jogging or walking and light stretching to help minimize residual soreness.

The mechanics of a horizontal sprint do not allow for easy quantification of the work performed, and power, the time derivative of work, cannot be measured. Therefore, time to complete the sprint is used for comparison purposes among different individuals. When speed is being used as a predictor of sport performance, it is usually best to compare times within a specific sport of interest. For example, the 40-yard dash has come to be a widely accepted predictor of football performance, and time for football athletes may not be useful in predicting success in other sports, such as basketball. There are relatively few norms for sprint tests. The American Alliance for Health, Physical Education, and Recreation (AAHPERD) has published normal values for the 50 yard dash for high school students, which are shown in Figure 2.1 (Hunsicker, Reiff, and American Alliance for Health, Physical Education, and Recreation, 1976).

FIGURE 2.1 — Fifty-yard dash times for 15 and 17+ year-old males and females (Modified from Hunsicker, Reiff, and American Alliance for Health, Physical Education, and Recreation, 1976).

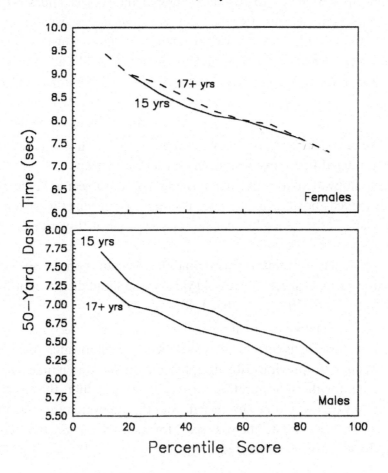

TWENTY-YARD SHUTTLE RUN (5-10-5 YARD SHUTTLE)

This is a test stressing the phosphagen system, which combines agility and speed in a task that requires quick, coordinated changes of direction, acceleration, and deceleration of the body. Many sport coaches consider this test as one of the most predictive of athletic success, especially in sports requiring agility and speed. For example, this shuttle run along with the 40-yard dash is a test used in the National Football League (NFL) combine draft camp to assess a player's athleticism and potential for a professional football career. Other shuttle distances may also be used in sport and fitness testing depending on the energy system to be challenged. As noted previously, maximal effort tests longer than about 12 to 15 seconds will begin to involve glycolytic processes in addition to the phosphagen system.

Equipment Needed

Same as for the anaerobic sprint test above

Test Preparation

1. Test and subject preparation is essentially the same as for the anaerobic sprint test above, except for the course layout, as described below.
2. Use a measuring tape to accurately place three marking cones and marking tape 5 yards apart in a straight line on the running surface. The course set-up is illustrated in Figure 2.2.

Test Procedure

1. Upon arrival, lead the subject through a series of stretching and jogging/running warm-up activities, gradually increasing in intensity and carefully designed to prepare the subject for this all-out shuttle task. As noted previously, the warm-up activities preceding any physical test should be standardized so that all subjects receive the equivalent warm-up, and the warm-up activity is the same each time the test procedures are repeated over the course of training.
2. Thoroughly explain and demonstrate the test procedure to the subject. Walk through the course showing the subject that the test is a 5-yard sprint in one direction, a 10-yard sprint in the opposite direction, and finally a 5-yard sprint to the start/finish line.
3. The subject should be asked to straddle the start/finish middle line, feet about shoulder width apart, and assume a three-point stance, with one hand on the ground as the start position (Figure 2.3).
   a. The timer will be stationed in front of the middle cone, facing the subject. The instructions to the subject are "On Your Mark" to call the subject to the starting line, then "Get Set" to bring the

FIGURE 2.2 — Illustration of 20-yard shuttle configuration.

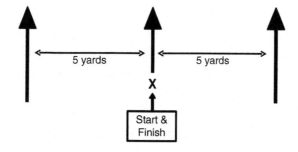

FIGURE 2.3 — Subject in position to start the 20-yard shuttle.

   subject to the three-point stance-ready position. The subject starts at their discretion, and timing begins on the subject's first movement after they assume the "set" position. Alternately, the timer can give a "Go" command after the subject is set.
   b. The subject is instructed to start in either direction according to their preference, and negotiate the course as fast as possible.
      i. Run 5 yards and touch the line with the hand corresponding to the start direction (e.g., right hand if started to the right).
      ii. Turn and sprint full speed 10 yards in the opposite direction past the start line and touching the 10-yard mark with the opposite hand from that used to touch the first line (e.g., left hand if start direction was to the right).
      iii. Turn again as fast as possible and sprint 5 yards *through* the start/finish line.
   c. The timer stops the stopwatch when any part of the runner's body crosses the plane of the start/finish line at the completion of the 5-10-5 yard sequence.
   d. The player is required to touch the line at each turn, or the effort is not counted.
   e. It is common practice to give the subject three attempts at this task allowing sufficient time for complete recovery between each attempt.
4. Record the times for the efforts in the appropriate space on the Speed, Agility, and Short-Term Anaerobic Power Data Sheet. The fastest time is typically used as the criterion score.
5. After the final test, direct the subject in cool-down routines, including slow jogging or walking and light stretching to help minimize residual soreness.

VERTICAL JUMP-AND-REACH TEST

This anaerobic power test, which again taxes the phosphagen system, is quite simple to administer, and is one of the tests almost universally applied to testing athletes at all levels and in all sports. It is often simply referred to as Vertical Jump Test. As with the 40-yard dash and the 20-yard shuttle, it is an important component of the test battery administered to prospects attending NFL combine draft camps. First described by Sargent in 1921 as the jump-reach test, it simply measures a person's vertical jumping ability from a flat-footed standing start. The unit of measure in the United States is usually inches, and the actual vertical jump is calculated as Maximal Jump-and-Reach Height minus the Standing Reach Height. This test has been criticized because the simple measurement of vertical jumping distance does not allow for a direct comparison of anaerobic leg power between individuals, unless those being compared are of equal body weights (remember that *power = force × distance ÷ time*). However, using the data available from the vertical jump test, several methods have been described to estimate anaerobic power. One of the most popular is the Lewis formula (Gray, Start, and Glencross, 1962). The logic for this formula is as follows. Body weight is a measure of the force of gravity acting on the body mass, and the distance over which the force acts can be measured as the vertical height the center of gravity of the body is lifted. Using a rate constant for a falling body due to the force of gravity, the vertical jump, and the body weight, leg muscle power can be calculated as follows. (Gray et al., 1962). Note that body weight must be in kilograms and the jump-height in meters.

$$\text{Power}_{(kpm \cdot sec^{-1})} = 2.21 \times \text{Body Weight}_{(kg)} \times \sqrt{\text{Jump Height}_{(m)}}$$

$$\text{Power (W)} = \text{Power}_{(kpm \cdot sec^{-1})} \div 0.1020$$

Equipment Needed

Vertec® Jump Measurement System or chalk or other marker

Measuring tape

Paper (2 ft × 5 ft)

Vertical surface such as a wall

Medical scale to measure height and weight

Test Preparation

1. Instruct the subject prior to arrival to wear comfortable exercise clothing and shoes for testing. No food should be consumed for at least 2 hours before reporting for testing, and drinks should be restricted to water only.
2. Set up the test area.
   a. <u>Preferable test equipment.</u> Set up the Vertec® Jump Measurement System on a level, firm surface with sufficient ceiling height to accommodate expected jump reach of the group to be tested (Figure 2.4). This measurement system is equipped with an adjustable height (6 to 12 feet) column of horizontal measurement vanes calibrated ½ inch apart, which are rotated aside by the hand of the test subject to indicate the height reached.
      i. Be sure the equipment is level. Check the calibration of the measurement scale on the instrument for accuracy with a measurement tape. Make calibration adjustments as needed according to the manufacturer's recommendations.
   b. <u>Alternate set-up.</u> Identify a flat, firm surface near a vertical wall with sufficient ceiling height to accommodate the expected vertical jump height of all in the group to be tested. Tape to a wall a piece of paper

FIGURE 2.4 — Standing reach measurement on Vertec® Jump Measurement System.

or cardboard large enough to span the height range of the standing reach and vertical jump reach of the shortest and tallest test subjects in the group.

   i. Have on hand markers or chalk for the subjects to use to mark their standing reach and vertical jump reach on the paper.

Test Procedure

1. If not already done in a previous testing step, the technician should measure the subject's height to the nearest 0.5 cm and weight to the nearest 0.25 kg, including any clothing that will be worn during the test. Record these data on the Speed, Agility, and Short-Term Anaerobic Power Data Sheet.
2. Provide directed warm-up activities to prepare for the vertical jump-and-reach task. As for all physical tests, the warm-up should be standardized between repeated measures occurring weeks or months apart so that serial measures of vertical jump are not affected by this factor.
3. Vertec® procedure.
   a. The test technician should obtain the subject's standing-reach height (SRH).
      i. Set the arm holding the column of measurement vanes low enough to be in reach of the subject from a standing position.
      ii. Instruct the subject to stand erect under the Vertec® measurement vanes, the shoulder of the dominant hand closest to the support pole, feet flat on the floor and close together, knees locked straight, eyes straight ahead. From this standing position, extend the dominant arm from the shoulder, locking the elbow straight upward, and stretch to reach as high as possible to rotate the measurement vanes at the highest reach (Figure 2.4)
      iii. Repeat at least three times, or until there is no further increase, then record the highest value in the SRH space on the Speed, Agility, and Short-Term Anaerobic Power Data Sheet.
   b. Obtain jump-reach height (JRH).
      i. Instruct the subject to assume the same standing position under the Vertec® measurement vanes as with the standing reach, except the feet should now be shoulder-width apart and the knees slightly flexed.
      ii. At will, the subject should in one jumping motion drop the hips bending at the knees as a counter movement, then jump as high as possible taking off from both feet (no step allowed), and reaching as high as possible with one arm to rotate the measurement vanes with the fingertips at the highest point of the jump-reach (Figure 2.5). This will be recorded as the JRH.
      iii. Repeat at least three trials, or until there is no further increase in the JRH. Record the values of each JRH trial in the appropriate spaces on the Speed, Agility, and Short-Term Anaerobic Power Data Sheet.
4. Alternate paper mark procedure.
   a. Mark and measure the SRH.
      i. The technician instructs the subject to hold the marker with the tip level with the end of the fingers on the dominant hand.
      ii. Stand erect with the shoulder of the dominant hand close to touching the wall, feet together and flat on the floor.

FIGURE 2.5 — Jump-reach height measure on Vertec® Jump Measurement System.

iii. Extend the arm and hand from the shoulder to place a mark on the paper at the highest point possible while still remaining flat footed.
iv. Ask the subject to step away from the paper, then move back into the standing position, and repeat the standing reach procedure. The two marks should be within 1 cm of each other. If they are not, repeat the procedure.
v. Measure and record the highest value in the SRH space on the Speed, Agility, and Short-Term Anaerobic Power Data Sheet.

b. Mark and measure the JRH.
   i. Instruct the subject to assume the same position as the standing reach, except the feet should now be shoulder-width apart and the knees slightly flexed.
   ii. At any self-selected time the subject should in one jumping motion drop the hips bending at the knees as a counter movement, then jump as high as possible taking off from both feet (no step allowed) and extend the arm and hand with the marker as high as possible making a mark on the paper at the peak of the jump-reach.
   iii. Repeat at least three trials, or until there is no further increase in the JRH.
   iv. Measure the jump-reach trials and record the values in the JRH spaces on the Speed, Agility, and Short-Term Anaerobic Power Data Sheet.

c. Calculate the vertical jump (VJ) as the difference between JRH and SRH. Convert VJ to meters for the muscle power calculation using the conversion; 1 inch = 0.0254 meter.

5. Calculate muscle power on the Speed, Agility, and Short-Term Anaerobic Power Data Sheet using the Lewis equation. (Gray et al., 1962).

$$\text{Power}_{(kpm \cdot sec^{-1})} = 2.21 \times \text{Body Wt}_{(kg)} \times \sqrt{\text{Vertical Jump}_{(m)}}$$

$$\text{Power}_{(W)} = \text{Power}_{(kpm \cdot sec^{-1})} \div 0.1020$$

STANDING LONG JUMP

Less commonly used than the jump-and-reach test, the standing long jump is another test of explosive leg power that taxes the phosphagen system and is sometimes applied in sports settings, including in the NFL draft combine. This test is very simple to administer with minimal equipment requirements. One major drawback is that it does not easily permit the computation of power to facilitate comparisons between people of different body height and weight. The comparative measure is the distance jumped in meters or feet.

Equipment Needed

Measuring tape

Firm, nonslip takeoff area and soft landing surface (landing mats may be used for this purpose)

Line, board, or marking tape for the takeoff line

Test Preparation

1. This test will require a laboratory team member to observe the takeoff mark to judge a "scratch jump," one starting illegally over the takeoff line. Another team member will be needed to mark the landing and assist in measurement of the long jump.

2. Mark the takeoff line a sufficient distance from the landing area to accommodate a safe landing for all participants.
3. Communicate to the subject prior to arrival to wear comfortable exercise clothing and shoes for testing. No food should be consumed for at least 2 hours before reporting for testing, and drinks should be restricted to water only.

Test Procedure

1. One laboratory team member should lead the subject through an adequate warm-up, if this has not previously been done.
2. For the test, the subject stands with both feet on the takeoff line or board, about shoulder width apart. No part of the feet may extend over the line at any time prior to the jump takeoff. No steps are allowed, but the subject should be encouraged to bend the knees in preparation for the long jump and use arm and leg countermovements to generate forward and upward drive to maximize longitudinal distance traveled.
3. The subject attempts to jump as far out into the landing area as possible and land on both feet without falling or stepping backward.
4. The laboratory team member spotting the landing must mark the most rearward spot of any part of the subject's body to touch the landing surface. This means, for example, that if the subject lands feet first but uses a hand to reach back to steady him- or herself, the measurement point would be from the hand.
5. The two team members measure the distance in meters or feet from the starting line to the marked spot in the landing area.
6. Usually at least three attempts are given, and the longest jump distance is counted as the subject's score.
7. Record the attempts in the appropriate space on the Speed, Agility, and Short-Term Anaerobic Power Data Sheet.

STAIR-STEP TEST

Sometimes referred to as the Margaria-Kalamen test, this assessment of maximal anaerobic power of the phosphagen system requires stair climbing at maximal speed (Kalamen, 1968; Margaria, Aghemo, and Rovelli, 1966). In brief, the time it takes to sprint up a series of steps from the third to the ninth step in a stairway is measured with electronic switch plates embedded in rubber mats placed on the steps. The power produced is calculated from the time to climb the measured 6 steps, the vertical distance climbed in the 6 steps, and the weight of the subject. The duration of the test is usually under 3 seconds, so again this test may be used to test the maximal capacity of the phosphagen system. However, the test is seldom used today in sports or other testing settings. It requires an open stairwell of sufficient height clear of obstructions and with adequate run-up space. To perform the test properly requires practice, and there is a reasonably high risk of slips, trips, falls, and injury in the process of exerting maximal speed to climb stairs three at a time. The test is typically limited to young, healthy, coordinated individuals and is not recommended for testing older adults. Even athletic coaches may find the risk of injury outweighs the potential benefits when other, safer means of testing muscle power (e.g., jump-reach) are readily available. If this test is used, it is recommended that reasonable precautions be taken when asking subjects to perform it, and that sufficient practice be given prior to the test to lessen the risk of slips, trips, and falls.

Equipment Needed

Stairway with at least 10 meters run-up space and nine steps high

Automatic timing device accurate to 0.01 second

Measuring tape or meter stick

Medical scale to measure height and weight

Test Preparation

1. Instruct the subject prior to arrival to wear comfortable exercise clothing and shoes for testing. No food should be consumed for at least 2 hours before reporting for testing, and drinks should be restricted to water only.
2. Again, the test procedures are best managed by a team of at least three students, including the subject.
3. Set up the testing equipment prior to the arrival of the subject. The main equipment for timing the sprint up the stairs consists of two rubber mats with embedded switch plates to start and stop a timing device capable of recording time to 1/100 second. The mats need to be taped down securely on the third and ninth steps of a stairway. Connect the wires for the mat switches to the timer, taking care that no trip or fall hazards are left to interfere with the subjects' sprinting up the stairs. It is best to tape the wires securely out of the way.

    a. Check the system by making a run-through to be sure the timer is in working order.

4. Measure the vertical height of steps 3 to 9, convert to meters (1 inch = 0.0254 m) then sum the heights of these 6 steps and record the summed value in meters as the vertical distance climbed (VDC) on the Speed, Agility, and Short-Term Anaerobic Power Data sheet.
5. Measure a distance 6 meters from the first step and place a line at this distance. This will constitute the starting line for the run up to the steps.

Test Procedure

1. If not done previously in completing other anaerobic power tests, obtain the subject's height to the nearest 0.5 cm and body weight to the nearest 0.25 kg. The subject should do this wearing the clothing that he or she will be wearing during the test. Record the weight and height on the Speed, Agility, and Short-Term Anaerobic Power Data Sheet.
2. As with all other anaerobic power test procedures, one laboratory team member should lead the subject through an appropriate warm-up before the test is attempted. The warm-up should include at least two practice runs at moderate speeds up the stairs before attempting the maximal sprint stair-step test. Instruct the subject to sprint up three steps at a time, being sure to strike only steps 3, 6, and 9, especially striking the center of the timing mats on steps 3 and 9.
3. One laboratory team member should check to reset the timer, and ask the subject to start behind the 6 meter line. Once behind the line, they may start when desired running as rapidly as possible to the steps, then climbing the stairway 3 steps at a time as fast as possible.
4. The timer will start when the subject steps on the timing mat on the third step and stops when the subject steps on the mat's ninth step.
5. It is common to give the subject at least three trials, or until there is no significant improvement in the time, with complete recovery between each. Record each trial time on the Speed, Agility, and Short-Term Anaerobic Power Data Sheet.

6. Calculate the power in both kpm · sec^{-1} and in watts for each of the trials performed using the formula on the Speed, Agility, and Short-Term Anaerobic Power Data Sheet, and also shown here.

$$\text{Power}_{(kpm \cdot sec^{-1})} = \text{Body Weight}_{(kg)} \times \text{Vertical Distance Climbed}_{(m)} \div \text{Time}_{(sec)}$$

Where: Body weight = the scale weight in kg of the subject
Vertical distance climbed (VDC) = Sum height in meters of 6 steps 3 to 9
Time = Time in seconds to climb from stairs 3 to 9

$$\text{Power}_{(W)} = \text{power}_{(kpm \cdot sec^{-1})} \div 0.1020$$

7. Choose the maximal stair-climbing power calculated, and record in the appropriate blanks on the data sheet.

PROCEDURES FOR MODERATE-TERM ANAEROBIC CAPACITY TESTS

The use of "moderate term anaerobic capacity" for our purposes refers to events or tests lasting longer than about 20 seconds, but less than 3 minutes. In exercise of this duration, the predominant pathway for energy formation after the phosphagen system is depleted in the first 10–15 seconds is anaerobic glycolysis. Events that exemplify the predominant use of the glycolytic system to generate the required ATP are the 400- and 800-m runs in track and field, and the 100- and 200-m races in swimming. Most of the tests of this system are very fatiguing, and thus can only be performed once unless ample rest is given between tests. Therefore, proper subject instruction is important for the validity of the tests. Two tests will be described in this section, one a simple step test that can be used in a mass testing situation, and the Wingate Cycle Ergometer test, more adapted to laboratory testing situations.

MANAHAN-GUTIN STEP TEST

This test is performed much like other step-tests described in the *Submaximal Exercise Testing for Estimating Aerobic Capacity* chapter of this laboratory manual, except it is an effort dependent timed test, in which power is calculated from the number of times an individual can step up and down on a step-box of known height in 1 minute (Manahan and Gutin, 1971). The results of this test have been shown to correlate well with a 600-yard run time in high school age girls, and test-retest reliability has been found to be high. However, further correlations with other standard running events have not been reported. The original test measured only the number of steps performed by the subjects in 1 minute (Manahan and Gutin, 1971). To that a computation of power can be added to obtain a better comparison with other anaerobic power measurements. Power output can be calculated using the following equation:

$$\text{Power}_{(kpm \cdot min^{-1})} = \text{Body weight}_{(kg)} \times \text{Number of steps in 1 min} \times \text{Bench height}_{(m)}$$

Where: Body weight = The scale weight of the subject in kg
Number of steps in 1 min = Total number of complete steps in 1 min
Bench height = Measured height of the bench in meters

$$\text{Power (W)} = \text{Power}_{(kpm \cdot min^{-1})} \div 6.12 \;_{(kpm \cdot min^{-1} \cdot W^{-1})}$$

Equipment Needed

18-inch step-box or bench
Medical scale to measure body weight and height
Stopwatch

Test Preparation

1. Instruct the subject prior to arrival to wear comfortable exercise clothing and shoes for testing. No food should be consumed for at least 2 hours before reporting for testing, and drinks should be restricted to water only.
2. Measure the step-box to confirm an 18-inch height.
3. Place the step-box or bench in such a way that the box will not slide when the subject steps on it.
4. It is most efficient for this test to be conducted in teams of three: one subject, one team member to time the test and communicate with the subject, and one to count the steps.

Test Procedure

1. If not done so for a previous test, weigh the subject in the clothing they will be wearing for the test to the 0.25 kg and height to the nearest 0.5 cm. Record these values appropriately on the Moderate-Term Anaerobic Power Data Sheet.
2. The technician should lead the subject in a brief warm-up in preparation for the test.
3. Provide the following instructions to the subject and demonstrate the movement while verbally giving instructions.
   a. The test will be 1 minute in duration. You will start on the "Go" command and need to make every effort to complete as many steps as possible over the test time.
   b. Your beginning position will be standing facing the step-box with one foot on the step-box and the opposite foot on the floor (Figure 2.6).
   c. For the laboratory team member to count the step, you must step up on the step-box to a full knee extension with the knee straight at the top of the step. It is not required that you actually step on the bench with your free foot being lifted, as long as this free foot reaches a level even with the top of the step-box with each step.
   d. You may alternate step-up legs as many times as you wish during the test.
4. Once the subject understands the procedures, ask them to assume the starting position. Provide them with the commands "Get set" followed by "Go." On the "Go" command, the subject should begin the step-test and the team member should simultaneously start the stopwatch.
   a. A second team member simultaneously begins to count only full steps, knee extended on top of the step-box and the free foot at least to the level of the top of the step-box.
   b. The team member timing the test should communicate encouragement to the subject and periodically announce the time remaining. Count out loud each of the last 10 seconds.
   c. The subject may elect to briefly rest during the timed minute of the test, but the timer does not stop the stopwatch during this rest.
5. At the end of the test, record the number of steps completed in the appropriate space on the Moderate-Term Anaerobic Power Data Sheet.
6. Instruct the subject in proper cool-down, including walking and stretching techniques.

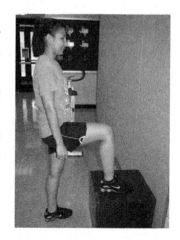

FIGURE 2.6 — Starting position for Manahan-Gutin step test.

7. Calculate the power output using the equation presented above and also shown on the Moderate-Term Anaerobic Power Data Sheet.

WINGATE CYCLE ERGOMETER TEST

The Wingate Cycle Ergometer test was developed to assess both short-term and moderate-term anaerobic capacities. It is the most widely used of the laboratory tests to measure moderate-term anaerobic capacity, but is not easily adaptable to field applications or to testing large numbers of individuals. Due to the intense nature of the test, it is also not advised for older populations and those who are very unfit. This standard test lasts for 30 seconds but may be modified to 15-second durations or less depending on the testing objectives. For example, football players in game situations exert maximal effort for about 6 seconds per play, and then recover between plays for about 30 seconds. In testing these athletes, it might be prudent to shorten the Wingate test to 10–12 seconds, or perhaps devise a protocol of repeat all-out 6-second bouts with 30-second rest periods between each. In this present laboratory exercise, the student will complete the full 30-second Wingate protocol. This 30-second version stresses the anaerobic phosphagen system in the first 10 to 15 seconds and the anaerobic glycolytic energy system thereafter. It is common to compute power for every 5-second segment of the 30-second test, so peak-cycling leg power can be measured, analogous to the maximal power calculated from the Sargent jump-reach test and the Margaria-Kalamen stair test. Average power over the 30 seconds, a fatigue index (peak power minus final 5-second power), and total work done over the 30 seconds are additional measures determined from this test. It is also typical to plot the 5-second data over the 30-second test to produce a "power curve," which can be compared between athletes or training periods to help guide training objectives to improve identified weakness. This illustrates the versatility of this test, since the power calculated the first 10–15 seconds provides an assessment of the maximal capacity of the phosphagen system, whereas power later in the test is a measure of the energy-producing capacity of anaerobic glycolysis. A number of investigations have shown a relatively high correlation between the results of the Wingate test and several other anaerobic power tests, as well as good test-retest reliability (Inbar and Baror, 1986). A number of automated cycle ergometer systems have been developed which are programmable for the Wingate test. These computerized systems make conducting the test and generating the data much easier and are highly recommended if Wingate tests will be performed routinely. The student is encouraged to use an automated system if available. In that case, the manufacturer's procedures will need to be followed to ensure data accuracy. However, any friction-braked cycle ergometer such as a Monark®, in which the resistance force and distance pedaled are calibrated and measurable, can be used. The student will use a manual method for the Wingate test in this laboratory.

Equipment Needed

Friction-braked cycle ergometer
Alternately, automated cycle ergometer system if available
Stopwatch
Medical scale to measure body weight and height

Test Preparation

1. Instruct the subject prior to arrival to wear comfortable exercise clothing and shoes for testing. This is an exhaustive test, so it is important that no food be consumed for at least 2 hours before reporting for testing, and drinks should be restricted to water only.

2. If an automated system is not being used, a testing team of three will be required: one subject, one team member to set the ergometer resistance, time the test, and communicate with the subject, and one to count pedal revolutions in each 5-second split over the 30-second test.
3. The testing team must be familiar with the operation of the equipment before the test begins. This is especially true if an automated system is used. Ensure the calibration of the cycle ergometer before subject arrival.

Test Procedure

1. If not already done in a prior test, measure body weight to the nearest 0.25 kg and height to the nearest 0.5 cm. This should be done with the subject wearing the clothing he or she will wear during the test. Record the body weight and height on the Moderate-Term Anaerobic Power Data Sheet.
2. As with all other anaerobic power test procedures, a testing team member should lead the subject through an appropriate warm-up before the test is attempted. The warm-up should include light cycling at about 60 rpm and a resistance of 0.5–1 kp for at least 3 minutes.
3. Calculate the resistance setting to be used for the test as shown on the Moderate-Term Anaerobic Power Data Sheet. The resistance setting, sometimes also called workload, is calculated by multiplying the subject's body weight by 75 gm · kg body weight^{-1}. As an example, if the subject weighs 80 kg, the load would be 6,000 gm or 6.0 kp (remember that 1 kg = 1 kp at sea level).
4. One team member should demonstrate to the subject the proper riding technique and pedal frequency before he or she mounts the ergometer. The demonstration should also include a verbal description communicating the following important points.
   a. This is a maximal cycling test in which you will pedal as fast as you can against the ergometer resistance for 30 seconds. Although it will be quite physically challenging, make every effort to complete the test. You may stop it if at anytime you feel you absolutely cannot continue. No rest period is permitted, so if you need to stop that will terminate the test.
   b. Once seated on the ergometer and ready to start the test, the commands will be "Get set," which means you should prepare to start pedaling as fast as you can, and then "Go," at which point you begin to pedal as fast as you can. You will be given approximately 5 sec to get the pedal rate up to your maximal speed. Then the team member will apply the resistance to the flywheel of the cycle ergometer.
   c. From the time you feel the resistance you will need to pedal as fast as you can for 30 seconds.
   d. The timer will verbally alert you to the passing of each 5-second period throughout the test.
5. Ask the subject to be seated on the ergometer, hands on handlebars, and adjust the ergometer seat so that there is a slight bend (15–25°) in the subject's leg when fully extended on the pedal. Strap the toes into the pedal toe straps of the cycle ergometer.
6. When the subject indicates a readiness to begin the test, the technician responsible for setting the resistance should give the "Get Set" and then "Go" commands. The subject then begins pedaling, gradually pedaling faster until the subject reaches maximal pedal velocity.
7. As soon as the subject reaches maximal velocity, and within 5 seconds, the technician should set the test resistance.
   a. At precisely the time-point when the resistance force has been set, start the stopwatch.
   b. At *exactly* the same time that the stopwatch is started, a second technician should begin counting pedal revolutions and keep track of the revolutions for each 5-second split. One way to do this is to

sit closely to observe the cycle pedals, and use a pen and paper to make a tick mark in a row across the paper for each complete pedal revolution. Start a new row of tick marks with each 5-second split, or use another differentiating mark on the paper to indicate the end and beginning of each 5-second split.

c. The team member timing the test must also loudly call out each 5-second split throughout the 30-second test and verbally encourage the subject to exert maximal effort through the full 30 seconds of the test.

8. At the end of 30 seconds of cycling at the test resistance:
   a. The timer loudly calls out the end of the test and immediately reduces the resistance to 1 kp and instructs the subject to slow the pedal rate for an active cool down. Encourage the subject to continue pedaling at a slow rate for 1 to 3 minutes for an active cool-down period.
   b. Stop counting pedal revolutions.

9. Record the number of pedal revolutions for each 5-second period in the designated blanks on the Moderate-Term Anaerobic Power Data Sheet.

10. The following text provides additional explanation for the calculations which are to be completed on the Moderate-Term Anaerobic Power Data Sheet. Note that automated cycle ergometer systems will make all calculations internally and print out a report summarizing all the data collected on the test. The report can usually be configured to print and plot the data most important for the athlete, coach, or trainer.

   a. First calculate the work done in each 5-second period. Use the formula below, which assumes the use of a Monark® friction-braked cycle ergometer. Other cycle ergometers may have a different distance constant for each pedal revolution, and this formula will need to be adjusted accordingly.

   $$\text{Work}_{(kpm)} = \text{Resistance}_{(kp)} \times \text{Number of pedal revolutions} \times 6 \text{ m} \cdot \text{pedal revolution}^{-1}$$

   where: resistance = kilopond (kp) of force
   Number of pedal revolutions = number of pedal revolutions in the 5-second period
   $6 \text{ m} \cdot \text{pedal revolution}^{-1}$ = constant 6 meter distance per 1 pedal revolution

   b. Calculate power in $\text{kpm} \cdot \text{sec}^{-1}$ for each 5-second split. This is accomplished by dividing the work done in kpm by 5 seconds, the duration of the split. Also convert each power to units of watts ($1 \text{ kpm} \cdot \text{sec}^{-1} = 9.81$ watts).

   c. Add the six 5-second work values together to obtain the Total Work Performed in 30 seconds as shown on the Moderate-Term Anaerobic Power Data Sheet.

   d. Calculate the average power output expressed per minute. To do this, multiply the total work done in 30 seconds by 2, to account for the fact that there are two 30-second periods in a minute. To convert power into units of watts (W), divide the power expressed in $\text{kpm} \cdot \text{min}^{-1}$ by $6.12 \text{ kpm} \cdot \text{min}^{-1} \cdot \text{W}^{-1}$. Also express the average power in $\text{kpm} \cdot \text{sec}^{-1}$ as shown in the Moderate-Term Anaerobic Power Data Sheet.

   e. Calculate the Fatigue Index on the data sheet, defined as the percent decrease in power output from peak power over the course of the 30-second test. This provides some insight into the anaerobic endurance capacity of the subject and can help target training weaknesses in anaerobic athletes.

   f. Many power test values are indexed on body weight to account for differences in body size. Calculate the Relative Average and Relative Peak Powers as shown on the data sheet.

QUESTIONS AND ACTIVITIES

1. Compare and contrast the power outputs from the short-term and moderate-term power. If there are differences, why might this be the case?
2. What factors other than the amount of ATP-PC could affect the short-term power on the tests performed in this lab?
3. How do the power values for the stair test, the jump reach test, and the first 5 seconds of the Wingate Cycle Ergometer test compare? Can you give a reason for the differences, if any?
4. Choose an anaerobic test or tests to apply to the following athletes, and give reasons why the information would be useful in developing targeted training programs to improve performance.
   a. Football lineman
   b. Softball player
   c. Woman soccer player
   d. Marathon distance runner

QUESTIONS AND ACTIVITIES FOR GOING FURTHER

1. Plot the split velocities for the anaerobic sprint test, and evaluate for the maximal velocity, when it was attained, and the percent of deceleration thereafter. Discuss how this information would help direct training to improve sprint speed?
2. Choose any of the test procedures for measuring short-term anaerobic power. What would you predict would be the effects of ingesting creatine supplements on the test measures before and after the supplement if all training remained the same? Provide a metabolic rationale for your answer.
3. Use a computer program to plot the power curve from the Wingate Cycle Ergometer test from two subjects measured in this lab but with distinctly different training histories (e.g., one endurance trained and one sprint or resistance trained). Analyze the power curves to differentiate the effects of their training on this test of anaerobic power. Compare and contrast the findings related to peak power, average power, fatigue index, and total work done. Index the variables on body weight and discuss the results.

REFERENCES

American College of Sports Medicine. *Exercise Is Medicine*. Accessed March 18, 2011, http://exerciseismedicine.org/.

Gray, R. K., K. B. Start, and D. J. Glencross. "A Test of Leg Power." *Res. Q.* 33.1 (1962): 44–50.

Hunsicker P. A., G. G. Reiff, and American Alliance for Health, Physical Education, and Recreation. *AAHPER Youth Fitness Test Manual*. Washington: American Alliance for Health, Physical Education, and Recreation, 1976.

Inbar, O., and O. Baror. "Anaerobic Characteristics in Male-Children and Adolescents." *Med. Sci. Sports Exerc.* 18.3 (1986): 264–69.

Kalamen, J. "Measurement of Maximum Muscular Power in Man." Doctoral thesis, The Ohio State University, 1968.

Manahan, J. E., and B. Gutin. "One-Minute Step Test as a Measure of 600-Yard Run Performance." *Res. Q.* 42.2 (1971): 173–77.

Margaria, R., P. Aghemo, and E. Rovelli. "Measurement of Muscular Power (Anaerobic) in Man." *J. Appl. Physiol.* 21.5 (1966): 1662–64.

Powers, S. K., and E. T. Howley. *Exercise Physiology: Theory and Application to Fitness and Performance.* New York: McGraw-Hill, 2009.

Rawson, E. S., and J. S. Volek. "Effects of Creatine Supplementation and Resistance Training on Muscle Strength and Weightlifting Performance." *J. Strength Cond. Res.* 17.4 (2003): 822–31.

Sargent, D. A. "The Physical Test of a Man." *School and Society* 13.318 (1921): 128–35.

SPEED, AGILITY, AND SHORT-TERM ANAEROBIC POWER DATA SHEET

Subject Name: _____   Date: _____ / _____ / _____

Weight: _____ lb _____ kg   Subject Height: _____ in _____ kg

Anaerobic Sprint Test

Sprint distance: _____ (yd)
Trial 1 finish time (sec) _____

Trial 1 Split	Time (sec)	Duration (sec)	Velocity (yds · sec-1)
0–10	_____	_____	_____
10–20	_____	_____	_____
20–30	_____	_____	_____
30–40	_____	_____	_____

Trial 2 finish time (sec) _____

Trial 2 Split	Time (sec)	Duration (sec)	Velocity (yds · sec-1)
0–10	_____	_____	_____
10–20	_____	_____	_____
20–30	_____	_____	_____
30–40	_____	_____	_____

Trial 3 finish time (sec) _____

Trial 3 Split	Time (sec)	Duration (sec)	Velocity (yds · sec-1)
0–10	_____	_____	_____
10–20	_____	_____	_____
20–30	_____	_____	_____
30–40	_____	_____	_____

20-Yard Shuttle Run (5-10-5 Yard Shuttle)

Trial Times $_{(sec)}$

Trial 1_____ Trial 2_____ Trial 3_____ Fastest_____

Vertical Jump-and-Reach Test

Standing-Reach Height (SRH) _____(in) _____(m) (1 inch = 0.0254 m)

Vertical Jump (VJ) $_{(in)}$ = Jump-Reach Height (JRH) $_{(in)}$ − Standing-Reach Height (SRH) $_{(in)}$

	JRH $_{(in)}$	VJ $_{(in)}$	VJ $_{(m)}$	√VJ $_{(m)}$	*Power $_{(kpm \cdot sec^{-1})}$	*Power $_{(W)}$
Trial 1:						
Trial 2:						
Trial 3:						
Trial 4:						

*Power $_{(kpm \cdot sec^{-1})}$ = 2.21 × Body Wt $_{(kg)}$ × √Jump Height $_{(m)}$
*Power $_{(W)}$ = 0.1020 × Power $_{(kpm \cdot sec^{-1})}$

Greatest Jump-and-Reach: Inches/m _____/_____ Power _____$_{(kpm \cdot min^{-1})}$ _____(W)

Standing Long Jump

Long Jump Trials in Feet (') and inches (")

Trial 1_____ Trial 2_____ Trial 3_____ Longest_____

Long Jump Trials in Meters

Trial 1_____ Trial 2_____ Trial 3_____ Longest_____

Margaria-Kalamen Stair Test

Vertical Distance Climbed (VDC) = Sum of height for steps 3 to 9 in meters = _____ (m)

	Trial Time $_{(sec)}$	*Power $_{(kpm \cdot sec^{-1})}$	*Power $_{(W)}$
Trial 1:_____			
Trial 2:_____			
Trial 3:_____			
Trial 4:_____			

*Power $_{(kpm \cdot sec^{-1})}$ = Body weight $_{(kg)}$ × VDC $_{(m)}$ ÷ Time $_{(sec)}$
*Power $_{(W)}$ = 0.1020 × Power $_{(kpm \cdot sec^{-1})}$

Maximal stair climbing power _____$_{(kpm \cdot min^{-1})}$ _____(W)

MODERATE-TERM ANAEROBIC POWER DATA SHEET

Subject Name: _____ Date: ____/____/____

Weight: _____ lb _____ kg Subject Height: _____ in _____ kg

Manahan-Gutin Step Test

Number of steps completed: _____ Bench height (m): _____

<u>Calculate Power</u>

_____ = _____ × _____ × _____

Power $_{(kpm \cdot min^{-1})}$ = Body weight $_{(kg)}$ × Number of Steps in 1 min × bench height $_{(m)}$

Power $_{(W)}$ = Power $_{(kpm \cdot min^{-1})}$ ÷ 6.12 $_{(kpm \cdot min^{-1} \cdot W^{-1})}$ = _____ (W)

Wingate Cycle Ergometer Test

Calculate Resistance Setting for Subject

Resistance $_{(kp)}$ = Weight $_{(kg)}$ _____ × 75 $_{gm \cdot kg\,body\,weight^{-1}}$ = _____ g ÷ 1000 = _____ kg or kp

<u>5-Second Split Data</u>

5-Sec Split	Pedal Revolutions	5-Sec Work $_{(kpm)}$	5-Sec Power $_{(kpm \cdot sec^{-1})}$	5-Sec Power $_{(W)}$
0–5	_____	_____	_____	_____
5–10	_____	_____	_____	_____
10–15	_____	_____	_____	_____
15–20	_____	_____	_____	_____
20–25	_____	_____	_____	_____
25–30	_____	_____	_____	_____

Work $_{(kpm)}$ = Resistance $_{(kp)}$ × Number of pedal revolutions × 6 $_{m \cdot pedal\,revolution^{-1}}$

Power $_{(kpm \cdot sec^{-1})}$ = Work $_{(kpm)}$ ÷ Time $_{(sec)}$ (1 kpm · sec^{-1} = 9.81 W)

Total Work Performed in 30 seconds $_{(kpm)}$ = Sum of all 5-second work periods = _____ (kpm)

Average Power $_{(kpm \cdot min^{-1})}$ = Total work in 30 seconds × 2 = _____ (kpm · min^{-1})

Average Power $_{(W)}$ = Average Power $_{(kpm \cdot min^{-1})}$ ÷ 6.12 = _____ (W)

Average Power $_{(kpm \cdot sec^{-1})}$ = Average Power $_{(kpm \cdot min^{-1})}$ ÷ 60 $_{(sec)}$ = _____ (kpm · sec^{-1})

Peak Power = Highest 5-sec Power _____ (kpm · sec⁻¹) or _____ (W)

Fatigue Index (%) = (Peak Power − Final 5-Second Power) ÷ Peak Power × 100 = _____%

Power Values Relative to Body Weight

Relative Average Power $_{kpm \cdot kg^{-1} \cdot sec^{-1}}$ = Average Power ÷ Weight (kg) = _____ kpm· kg⁻¹· sec⁻¹

Relative Peak Power $_{kpm \cdot kg^{-1} \cdot sec^{-1}}$ = Peak Power ÷ Weight (kg) = _____ kpm· kg⁻¹· sec⁻¹

CHAPTER 3

TESTING OF MUSCLE STRENGTH, ENDURANCE AND FLEXIBILITY

OUTCOME OBJECTIVES

Following successful completion of this exercise, the student should be able to:

1. Define muscle strength, endurance, and flexibility, understand the reasons for testing these variables, and the bases for the tests performed.
2. Administer and interpret common tests of muscle strength, endurance, and flexibility.

INTRODUCTION

A very important part of whole body fitness is that of fitness of the musculoskeletal system. As we age, musculoskeletal fitness becomes increasingly important for acts of daily living (ACSM, 2010). Typically this fitness can be divided into three portions: muscular strength, muscular endurance, and muscle and joint flexibility. Muscular strength refers to the ability of a person to generate maximal force using a muscle or group of muscles. Muscular endurance emphasizes the ability to maintain repeated contractions or force generation for a prolonged period of time. This force can be generated without movement (isometric), or through a range of motion (variously referred to as isotonic, dynamic, concentric, or eccentric). Characteristic tests of muscle strength and endurance for specific muscle groups can be performed under either isometric or isotonic conditions with nearly equal ease. Isometric testing is often simpler conceptually and easier to quantify, but has the disadvantage of only testing muscle strength or endurance at one specific muscle length, while strength or endurance in the real world is needed through a range of lengths. Isotonic testing involves movement through a wide range of muscle length, so it may better mimic everyday activities. But to obtain maximal efforts sometimes requires more than one attempt, since the testing generally uses only one specific force during an individual effort. The tests to be described in this exercise will encompass certain aspects of both isometric and isotonic contraction of different muscle groups. We will not attempt to provide an exhaustive battery of tests for all muscle groups that can be tested. For that, the reader is referred to one of a number of testing books listed in the additional readings.

Muscle and joint flexibility encompasses elements within the muscle itself, in the connective tissue and tendon, and in the joint capsule and ligament. Therefore, flexibility may be decreased by muscle events such as spasms or tears, loss or damage of connective tissue or ligament elasticity, or by joint irregularities (Nieman, 2011). Flexibility can further be divided into static and dynamic flexibility (Fox, Bowers, and Foss, 1993). Static flexibility is the range of motion of a joint. This is what we typically think of as flexibility and can be measured easily with a goniometer at a joint or estimated by measurements of stretching. Dynamic flexibility is the resistance to movement of a joint. This is difficult to measure and so has not been dealt with to a great extent. It may be important, however, in that if a joint has little dynamic flexibility, the joint is more difficult to open and close, and the muscles surrounding that joint have to do more work in order to create the movements desired. In this exercise, we will measure only static flexibility and will not concentrate on the measurement or estimation of dynamic flexibility.

All of the variables discussed here, with the possible exception of dynamic flexibility, are readily adaptable to training. While it is not the purpose of this manual to discuss training programs, improvements in strength, endurance, and flexibility of muscle groups and joints are readily apparent following appropriate training programs. Estimating this improvement is one aspect of training that is easily measured, and thus one of the principal reasons for the advent of muscle strength, endurance, and flexibility testing. The tests described in this chapter represent but a small portion of the tests devised for measuring muscle strength, endurance, and flexibility. The exercise is designed to provide instruction on how to perform several of the most widely used tests that are readily adapted to either the laboratory or the field.

MUSCLE STRENGTH TESTS

Any test of muscle strength needs to be one that is simple to perform, rather than requiring the subject to master a new skill. It also needs to be replicable because only a few efforts can be made prior to fatigue unless ample recovery is given. The two tests described here meet those requirements.

ISOMETRIC HANDGRIP TEST

This is one of the simplest and most common of the muscle strength tests available. It is widely used, so there are many reports with which to compare the data gathered. The reliability of the measurement is good, with reliability coefficients of 0.9 or higher (Beam and Adams, 2011), also making this a good test to compare training responses as well as population differences. This test has the disadvantage, as discussed earlier, of only testing the strength of the muscle group at one length, without movement. It is also specific to one, relatively small, muscle group. However, because of its widespread use, reliability and ease of use, the isometric handgrip test merits a description and demonstration. See Figure 3.1.

FIGURE 3.1 — Handgrip test.

Equipment Needed

Handgrip dynamometer

Scale for subject weight

TECHNIQUES AND PROCEDURES

1. Preparation for this test, as with many field type tests, simply consists of gathering the equipment and making sure that the handgrip dynamometer fits the hand of the subject. Adjust the size so that when the subject grips, the middle knuckle of the middle finger is bent to about 90°.
2. Measure the subject's body weight and record this number in the appropriate space on your data sheet.

3. The subject may either stand, with the upper arm perpendicular to the floor, or sit with the forearm resting on a bench or table. Record the position from which the subject works so that you have a basis for comparison with other tests. If standing, the subject may have the elbow bent to an angle of 90°.
4. Instruct the subject to exert a maximal handgrip. Record this number in your data sheet.
5. Repeat this procedure with the other hand until you have obtained at least three maximal handgrip efforts with each hand.
6. Add together the best effort from each hand and record this number. This is your raw score. If your dynamometer records force in pounds, divide the result by 2.2 to obtain the result in kilograms.
7. Divide the sum by the subject's body weight in kilograms. The result is the handgrip in kg/kg body weight. Record this number in your data sheet.
8. Compare the results to those in Figures 3.2 and 3.3 to find a percentile score. Record this number in your data sheet.

FIGURE 3.2 — Graph showing handgrip sum against percentile rank. To calculate your score, find the sum of handgrips (right + left) on the Y axis for your gender and age group, then draw a vertical line to the X axis and record the number. For other age groups, refer to the original study.

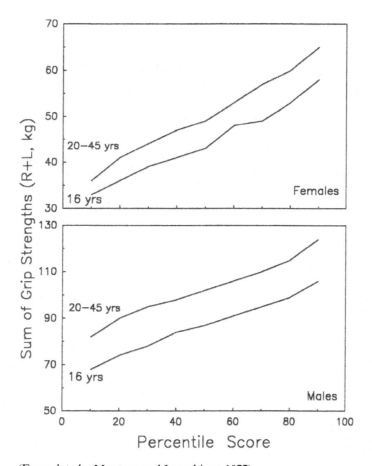

(From data by Montoye and Lamphiear, 1977)

FIGURE 3.3 — Graph showing the ranking for grip strength divided by body weight. Calculate the ranking as in Figure 3.1.

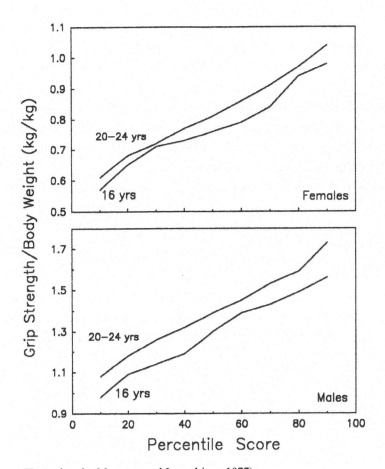

(From data by Montoye and Lamphiear, 1977).

BENCH PRESS MAXIMUM

The bench press is an example of an isotonic muscle strength measurement. The bench press primarily tests the strength of the arms, chest, and shoulders. Similar tests can be employed for other muscle groups if the same principles are applied to the testing protocol. To find some of these tests, the reader is referred to the additional readings. Since this test requires a maximal effort, the subject can seldom complete more than 3 to 5 lifts, and even then, ample rest is needed between efforts to ensure a successful test. Since this is a maximal test, care must be taken to prevent injury to the subject. See Figure 3.4.

Equipment Needed

Weight bench

Barbell and weights totaling 1.5X the weight of the subject, or weight machine such as

Universal Gym

Scale for subject weight

FIGURE 3.4 — Bench press in two positions.

© Netea Mircea Valentin, 2011. Used under license from Shutterstock, Inc.

© Netea Mircea Valentin, 2011. Used under license from Shutterstock, Inc.

TECHNIQUES AND PROCEDURES

1. Set up the test by putting weights on the bench or placing the bench at the weight machine.
2. Weigh the subject. Record this weight in your worksheet.
3. Set the initial weight for the lift. Often the subject will know how much weight to place; however, if the subject does not know, a reasonable starting point is 40% to 60% of the body weight.
4. Instruct the subject to lift the weight. A successful lift is one that goes from a position with the bar at the level of the chest to one with the arms fully extended. If a barbell with weights is used rather than a weight machine, it is imperative that two assistants be present to "spot" and prevent injury to the subject in the event of an unsuccessful lift.
5. Subsequent lifts will be lighter or heavier depending on whether the subject completed the previous lift. The amount of weight added or subtracted depends on the subjective impression of the subject's ease of completing the previous lift.
6. Record the weights successfully lifted. Do not have the subject try more than three to five attempts, as muscle fatigue will begin to be a factor in the subsequent attempts.
7. Divide the maximal weight lifted by the subject's body weight and record this number on your worksheet.
8. Compare this number to Figure 3.5 for category ratings.

MUSCULAR ENDURANCE TESTS

Tests for muscular endurance should not require a maximal effort on the part of the subject at the start of the test; this would place it in the category of a strength test. Instead, these should test the fatigability of a muscle or group of muscles. Most tests falling within this category require sustained efforts of one minute or less. However, they differ from the moderate term anaerobic capacity tests described in Chapter 2 in that they use a smaller group of muscles, thus eliminating most cardiovascular or other systemic input into the results of the test. As with the strength tests, these tests must be reproducible and reliable

FIGURE 3.5 — Graph showing bench press categories for college-age males and females. To determine the ranking, find the maximal bench press/body wt on the X axis and draw a vertical line until it reaches a bar. This is the category ranking.

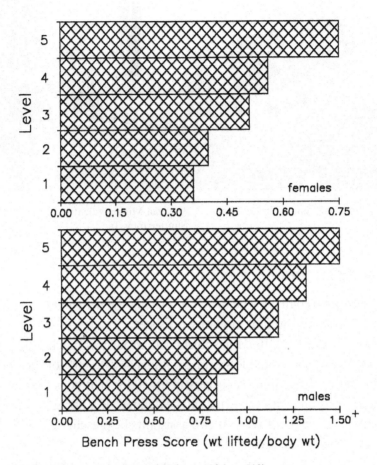

(Adapted from Johnson and Nelson, 1986, p. 113)

since they cannot be repeated even once without ample recovery. One other problem associated with muscular endurance tests is the motivation of the subjects. Since severe muscle fatigue comes into play in these tests it is important for the investigator to provide the same amount of encouragement to each subject and avoid the complications associated with extraneous motivational signals (Gettman, 1988).

FLEXED-LEG SIT-UPS

This is one of the few muscular endurance tests for which standards and rankings have been established for both males and females. Other tests are either standardized only for males or females (e.g., pull-ups or flexed arm hang) or are modified for testing males or females (e.g., push-ups or pull-ups) (Johnson and Nelson, 1986). For that reason the sit-up test will be described exclusively. This test assesses the endurance of the abdominal muscles and the hip flexors. Its use may be applicable to healthy lifestyle situations (in addition to strictly fitness assessment) because weak abdominal muscles are often associated with lower back pain in adults. See Figure 3.6.

FIGURE 3.6 — Sit-ups in two positions.

© Benjamin Thorn, 2011. Used under license from Shutterstock, Inc.

© Benjamin Thorn, 2011. Used under license from Shutterstock, Inc.

Equipment Needed

Timer

Exercise mat

TECHNIQUES AND PROCEDURES

1. The site for testing can be any place with a flat surface. It is best to provide a mat on which the subject can lie or to perform the test in a grassy area. Sit-ups done on a hard surface can lead to abrasions or bruising.
2. Have the subject lie in the supine position with the knees bent so that the subject's heels are approximately 18 inches from the buttocks. The arms should be crossed in front of the chest with the hands on the opposite shoulders. Alternatively, the hands can be placed by the side of the head. They should not be clasped behind the neck to prevent possible injury.
3. A successful sit-up consists of rising from the supine to the seated position with the elbows toughing the thighs. The investigator or an assistant may hold the subject's feet to the floor during the test.
4. Instruct the subject that he or she is to perform as many sit-ups as possible during a 1-minute period and that you will keep the subject informed as to the amount of time remaining.
5. From 5 seconds, count down the time and instruct the subject to begin doing sit-ups. Count each effort that meets the conditions set forth in number 3 above as one sit-up.
6. Each 15 seconds, inform the subject of the time remaining.
7. At the end of the 1-minute period, tell the subject to stop. In the appropriate space in the worksheet, record the number of successful attempts performed during the test.
8. Compare the number of sit-ups performed with the graph in Figure 3.7 to characterize the fitness of the subject.

JOINT FLEXIBILITY TESTS

Flexibility is widely recognized as both an asset to athletic performance and to health (Mathews, 1978), although excess joint flexibility may be detrimental to certain types of performances, such as contact sports in which shoulder flexibility may increase the likelihood of injury (Powers and Howley, 2009).

FIGURE 3.7 — Percentile rank for number of sit-ups for 17+ year old males and females. Calculate rankings as in figures 1 and 2.

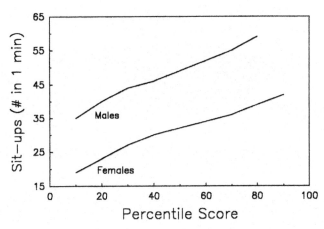

(Adapted from Tables 2 and 8, AAHPER, 1975)

Greater flexibility has been associated with decreased pain on exertion and decreased muscle soreness, although a specific cause and effect relationship has yet to be determined. As with strength and endurance, muscle and joint flexibility is responsive to training.

SIT AND REACH TEST

This test evaluates the flexibility in the hips and lower back. It is also responsive to the tightness of the hamstring and shoulder muscle groups. The sit and reach test is easy to perform, is reliable, and assesses flexibility in a relatively large group of muscles and joints. For that reason it is widely used in fitness settings and exercise science laboratories, often as the sole measure of flexibility, and will be the exclusive test described here. Other tests that work effectively have been described for other joints or limbs. For these the reader is referred to the additional readings.

Equipment Needed

   12-inch high box

   Yardstick

TECHNIQUES AND PROCEDURES

1. The measurement device for the sit and reach test can be purchased ready made, or can be made by attaching a yardstick to a box. The box should be sturdy and approximately 12 inches in height. The yardstick is attached to the box so that the 15-inch mark is at the edge of the box and the lower numbers extend past the edge.
2. The subject should be instructed and encouraged to participate in proper stretching and warm-up techniques prior to testing. These should include hamstring, back, and shoulder stretching.
3. To perform the test, have the subject seated with the legs extended so that the feet are placed flat against the edge of the box. The subject may either wear or not wear shoes. Since the heel height of most exercise shoes is rather small, any error induced will also be small.
4. Have the subject then place the hands in front of the body with index fingers together and reach the farthest point possible on the yardstick. The number used is the farthest reach that the subject can hold for 1 second.
5. Allow three attempts. A successful attempt is one in which the subject's legs were kept straight and in contact with the floor, and in which the subject did not bounce to attain the best distance.
6. Record the three attempts. Compare the best attempt with the graph in Figure 3.8 to categorize the performance of the subject. Record the category in the appropriate space in the worksheet.

FIGURE 3.8 — Category rankings for the sit and reach test for 20–29-year-old males and females. Calculate the ranking as in the bench press test.

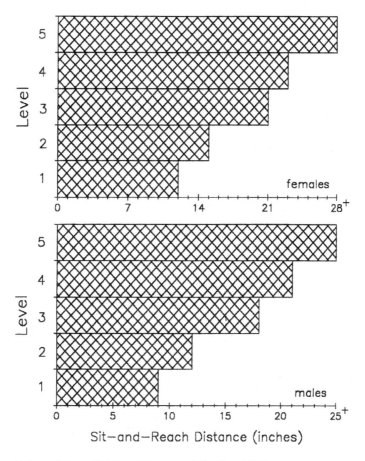

(Adapted from Golding, Myers, and Sinning, 1982)

QUESTIONS AND ACTIVITIES

1. Rank yourself as well as the average male group and average female group in the class on each of the tests administered.
2. How did you rank? How did the class rank? Would you expect these findings? Why?
3. What is the difference between muscle strength and endurance? Could we reliably measure muscle endurance in a one-repetition test, or muscle strength from a test that took 1 minute?
4. What are the advantages and disadvantages of isometric and dynamic means of testing muscle strength or endurance?
5. Why should bouncing not be allowed in the flexibility tests?
6. Provide a ranking for yourself as well as for the others being tested. Where do you rank? What about the mean of your group?
7. Is there a difference between males and females in your group? Can you suggest a reason why?

REFERENCES AND ADDITIONAL READINGS

American Association for Health, Physical Education, and Recreation. *AAHPER Youth Fitness Test Manual.* Washington: AAHPER: 1975.

American College of Sports Medicine. *ACSM's Guidelines for Exercise Testing and Prescription,* 8th ed. Philadelphia: Lippincott, Williams and Wilkins, 2010.

Beam, W. C. and G. M. Adams. *Exercise Physiology Laboratory Manual,* 6th ed. New York: McGraw Hill, 2011.

Fox, E. L., R. W. Bowers, and M. L. Foss. *The Physiological Basis for Exercise and Sport.* Dubuque, IA: Wm. C. Brown, 1993.

Gettman, L. R. "Fitness Testing." In: *Resource Manual for Guidelines for Exercise Testing and Presciption.* American College of Sports Medicine. Philadelphia: Lea and Febiger, 1988.

Golding, L. A., C. R. Myers, and W. E. Sinning, (eds.). *The Y's Way to Physical Fitness,* 3rd ed. Champaign, IL: Human Kinetics Publishers, 1989.

Johnson, B. L. and J. K. Nelson. *Practical Measurements for Evaluation in Physical Education.* Edina, MN: Burgess Publishing, 1986.

Mathews, D. K. *Measurement in Physical Education.* Philadelphia: W. B. Saunders, 1978.

Montoye, H. J. and D. E. Lamphiear. "Grip and Arm Strength in Males and Females, Ages 10 to 69." *Res Q.* 48 (1977): 109–20.

Nieman, D. *Exercise Testing and Prescription: A Health-Related Approach,* 7th ed. Boston: McGraw-Hill, 2011.

Powers, S. K. and E. T. Howley. *Exercise Physiology. Theory and Application to Fitness and Performance,* 7th ed. Boston: McGraw-Hill, 2009.

MUSCLE STRENGTH, ENDURANCE AND FLEXIBILITY WORKSHEET

Subject name: _____ Age: _____ Sex: _____

Height: _____ Weight: _____

Muscle Strength

Handgrip

Right Hand   Trial 1: _____   Trial 2: _____   Trial 3: _____   Trial 4: _____

Left Hand   Trial 1: _____   Trial 2: _____   Trial 3: _____   Trial 4: _____

Best Attempt

Right Hand: _____   Left Hand: _____   Sum (Rt + Lft): _____   Sum/Body Wt: _____

Percentile

Sum: _____   Sum/Body Wt: _____

Bench Press

Trial 1: _____   Trial 2: _____   Trial 3: _____   Trial 4: _____   Trial 5: _____

Best Attempt

Total Wt: _____   Total Wt/Body Wt: _____   Category: _____

Muscle Endurance

Flexed-leg Sit-ups

Number of sit-ups in one minute: _____   Percentile Rank: _____

Flexibility

Sit and Reach Test

Trial 1: _____   Trial 2: _____   Trial 3: _____

Best Attempt

Distance: _____   Category Rank: _____

PULMONARY FUNCTION TESTING

CHAPTER 4

OUTCOME OBJECTIVES

Following successful completion of this exercise, the student should be able to:

1. Be familiar with various commonly used pulmonary function measurements and their applicability to an exercise and a fitness setting.
2. Be able to conduct spirometry tests and maximal voluntary ventilation tests, measure residual volume, and calculate the results from these tests.
3. Be able to interpret the results of a basic spirometry test in light of a health or fitness situation.

INTRODUCTION

The measurement of pulmonary function is useful in the exercise physiology laboratory, particularly in those programs devoted to stress testing, health, and wellness, as well as those involved in pulmonary research. The various pulmonary function tests to be performed in this exercise measure volumes of air in the lungs and the rates at which these volumes can be inhaled or exhaled. Changes in these volumes and rates from normal values can be important indicators of health or disease, and thus exercise fitness. For instance, certain conditions such as asthma, bronchitis, and emphysema, which create obstructions to air flow within the body, exhibit characteristic changes in the spirometry test (Schwartzstein and Parker, 2006). Additionally, with the large number of people living in areas with high air-pollution levels, the acute and chronic effects of this pollution on the lungs and airways are becoming more prevalent and sometimes are seen in pulmonary function tests. Therefore, these measurements may be useful screening techniques for pulmonary disease, the effects of which may be slowed by early detection and treatment.

The spirometer is the principal piece of equipment used in most pulmonary function tests. If a subject simply breathes into and out of the spirometer, a number of characteristic lung volumes can be measured. The output from a spirometer is termed a spirogram (see Figure 4.1). A brief description of the lung volumes that an operator might see in a spirogram and the definitions of these volumes are as follows:

Tidal Volume (TV): The volume of air inhaled or exhaled in one normal breath.

Inspiratory Reserve Volume (IRV): The maximal amount of air that can be inhaled following a normal inhalation.

Expiratory Reserve Volume (ERV): The maximal volume of air that can be exhaled following a normal exhalation.

Inspiratory Capacity (IC): The maximal amount a subject can inhale following a normal exhalation.

FIGURE 4.1 — Spirogram during unforced breathing. Volume is on the Y axis, while time is on the X axis.

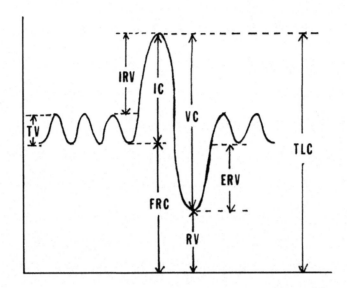

Vital Capacity (VC): The maximum amount of air that a subject can exhale after a maximal inhalation, or inhale following a maximal exhalation.

The following volumes or capacities cannot be measured with a spirometer, but require the use of more complicated dilution techniques or methods which simultaneously measure changes in lung volumes and pressures.

Residual Volume (RV): The amount of air that remains in the lungs after a maximal exhalation.
Functional Residual Capacity (FRC): The volume of air left in the lungs after a normal exhalation.
Total Lung Capacity (TLC): The total volume of the lungs.

Notice that some of the measurements use the term *volume* while others use *capacity*. There is a reason for this differentiation of terms. A volume is one non-overlapping measurement, while a capacity consists of two or more volumes added together. These volumes and capacities are generally used when characterizing subjects, and vary with body height, age, and gender, as well as with the pulmonary health of the subject. Residual volume is often used in the measurement of body composition since it is part of the volume of the body but carries no weight. In a healthy person, the above volumes are not indicators of fitness, and although most of the volumes change with disease states, these are seen only after the disease has progressed sufficiently to be detected by other, more precise, means. One of the other means to detect pulmonary disease is through the use of what are called forced volumes, which are maximal breaths performed using an all-out effort. Typical forced volumes and flows are as follows:

Forced Vital Capacity (FVC): The total volume expired after a maximal inhalation, during which time the subject attempts to exhale as rapidly and forcefully as possible. In a healthy subject, FVC should be the same as vital capacity above.

Forced Expiratory Volume in One Second (FEV$_{1.0}$): The amount of air exhaled in the first second of a forced vital capacity maneuver.

$FEV_{1.0} = \frac{actual}{predicted}$ > 80%, no OLD

$\frac{FEV_{1.0}}{FVC}$ > 70% indicates no Obstructive lung disease (OLD)

$FVC = \frac{actual}{predictive}$ > 80%, no Restrictive Lung Disease (RLD)

Forced Expiratory Flow from 25% to 75% of exhalation (FEF$_{25-75}$) or Maximal Mid-Expiratory Flow (MMEF): The flow rate during the middle 50% of the forced vital capacity maneuver, or from 25% to 75% of the exhaled volume.

The FVC and FEV$_{1.0}$ are the most commonly measured volumes and have been shown to be the best predictors of the presence or absence of disease in large test groups (Ferris, 1978). The FEF$_{25-75}$ is also described since it is a relatively common test and can be obtained from the forced expiratory spirogram used for the other forced measurements. Another typical measurement that can be obtained from the spirogram is the ratio between FVC and FEV$_{1.0}$, or the FEV$_{1.0}$%. This is calculated from the division of the FEV$_{1.0}$ by the FVC measurements and is widely used in the detection of disease. Other measurements can be made in certain instances during forced exhalations or inspirations if a physician determines the need to do so. However, these are generally tests to determine specific types or sites of disease and are not routinely performed in the exercise physiology laboratory.

Another relatively common and simple measurement is the maximal ventilatory volume (MVV). The MVV is the maximal amount of air that a person can breathe in or out in a short period of time, typically 10, 12, or 15 seconds. It is sometimes used as an indicator of disease, respiratory muscle weakness, or athletic ability, although its ability to predict the latter is not very strong unless there is prior indication of a pulmonary impairment. It is, however, commonly performed and may provide information about the health of the subject.

The major piece of equipment for the measurements to be made in this exercise is the spirometer. A spirometer measures volumes of air breathed in or out with respect to time, and a number of spirometers are available. These range from very simple drums with pens attached to fully automated computerized machines. They may measure volume directly, or measure flow and integrate the flow signal to obtain volume. Examples of different spirometers can be seen in Figure 4.2.

FIGURE 4.2 — Three commonly encountered types of spirometer. (*a*) Spirometer utilizing an inverted bell and attached recorder. (*b*) This spirometer employs a dry seal and has a pen attached. (*c*) A fully automated spirometry unit, which integrates flow into volume measurements and calculates results on a computer.

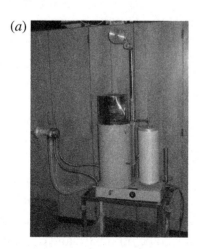
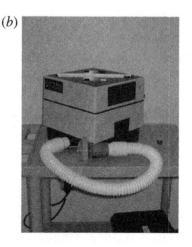

The spirometer to be used for this exercise consists of a lightweight bell, made of plastic or metal, inverted in a water-filled cylinder, and a recording device. This type of spirometer has been chosen because it is easy to understand conceptually, provides accurate results if used properly, and leads to a better understanding of the concepts needed to properly use the more complex machines.

Equipment Needed

Spirometer	Spirometer paper	Spirometer pens
Meteorological balloons	Ruler	Mouthpieces
Nose clips	Tubing	Disinfectant
Distilled water	Clear tape	Breathing valve
Oxygen analyzer	Gas volume meter	3-way stopcock valve

TECHNIQUES AND PROCEDURES

SPIROMETRY

1. Preparation of the spirometer: Fill the cylinder with *deionized* or *distilled* water to the fill level indicated by a clear opening on the side of the apparatus. It is important to use distilled or deionized water because most tap water contains a number of dissolved minerals that can lead to corrosion of the spirometer.
2. Attach clean tubing and a clean mouthpiece to the spirometer. These should be cleaned in disinfectant solution after each subject to avoid the spread of air- or saliva-borne diseases. If you have spirometer filters, a new one should be used for each subject.
3. Place the special ruled spirometer paper over the recording drum and tape it to the drum. Be sure that the lines match where the paper overlaps, or the measurement will not be accurate.
4. Insert the spirometer pen in the holder, making sure that the recording pen is working and is in contact with the recording paper.
5. Seat the subject comfortably in front of the spirometer with the back straight and the chin level or pointed slightly upward rather than resting on the chest. This posture helps keep the airways as open as possible.
6. Place the nose clips on the subject. During the procedures, the subject *must* wear a nose clip.
7. Set the recording drum at a slow rotating speed and instruct the subject to breathe normally through the mouthpiece 3 to 7 times. Then have the subject inhale maximally and breathe out as far as possible.
8. With the subject off of the mouthpiece, raise and lower the spirometer bell manually several times to flush the spirometer with fresh room air. Repeat the procedure from step 7. The output on the record should resemble the graph in Figure 4.1. The values for each of the volumes can be obtained by reading the difference in the volumes on the record. Record the values for each variable on your worksheet.
9. Forced volumes. The method for determining forced volumes is similar to that for the spirometric volumes. When the subject is seated comfortably with nose clip on, have the subject breathe normally for 2 or 3 breaths. Then turn the speed controller on the recorder to the fast speed and instruct the subject inhale maximally, followed by a maximal exhalation, which is to be both hard and complete. When performing the force expiratory tests, it is generally necessary to coach the subject.
    a. Make sure that the subject has his or her lips sealed around the mouthpiece, so no air leaks out on expiration.

b. Instruct the subject to keep the tongue out of the mouthpiece opening. This will interfere with airflow.
   c. Coach the subject to inhale as fully as possible then exhale as hard and fully as possible. You may say something such as, "OK, breathe in as much as you possibly can, keep going; NOW BLOW!, Keep going, keep going, more, more. Don't stop!"
10. The record you obtain should resemble part I of Figure 4.3. To be sure that your forced expiratory spirogram is acceptable, look out for several key characteristics.
   a. At the point at which the subject began exhaling, there should be a clear breakpoint from the spirogram baseline, where you can see that THIS is the point at which the exhalation began. If the point is too rounded, it indicates that the subject began exhaling slowly prior to the forced exhalation, and it is very difficult to determine at exactly which point the exhalation began. The spirogram is unacceptable.
   b. During the exhalation, is the curve smooth and rounded, or is there a "hitch" in it that would indicate a cough or variation in effort? If it is not smooth, the curve is unacceptable.
   c. Is there a plateau at the end of the spirogram? A plateau is defined as less than 25 ml change in ½ second, and looks like NO change in volume with continued expiratory effort. If such a plateau does not exist, it indicates that the subject did not exhale to residual volume, and you have not measured a full vital capacity. In a healthy subject this may take only 3 to 5 seconds, but in a subject with severe asthma it may take as much as 10 seconds or more.
   d. In addition to the acceptability of each spirogram as defined above, in order to have an acceptable test, the two best forced expiratory curves must have a vital capacity within 5% or 100 ml, whichever is greater.

FIGURE 4.3 — Three forced spirograms. I. This is an example of a spirogram which meets the criteria for acceptability. II. This spirogram shows (A) a rounded start, indicating an invalid point from which to measure the beginning of exhalation, and (B) a cough during the expiratory maneuver. III. This spirogram shows no plateau at the termination (C) and indicates the exhalation should have been carried out for a longer duration.

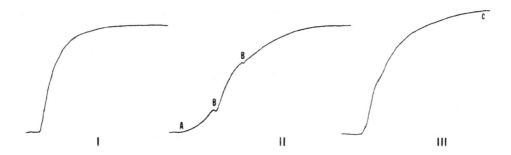

11. Following each exhalation, flush the spirometer with fresh air. Then repeat the maneuver until you have obtained three acceptable spirograms. A sample spirogram for use in practice calculations is shown in Figure 4.4.
12. Calculation of the FVC. The value for FVC is simply the difference between the reading of the volume recorded at the start of exhalation (the breakpoint) and that recorded from the end of exhalation (the plateau). These volumes can be read directly from the spirometer paper. In Figure 4.4, these

FIGURE 4.4 — Forced spirogram of approximately actual size. Use this spirogram for sample calculations, following the steps listed below.

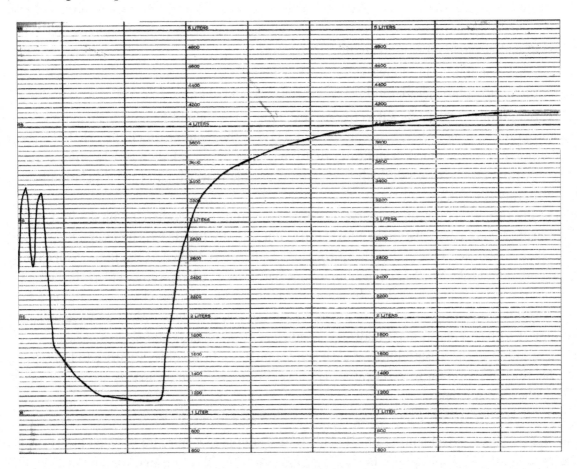

values are 1,150 ml and 4,130 ml, respectively, resulting in a difference of 2,980 ml, which is the FVC for this tracing.

13. Calculation of $FEV_{1.0}$. The value for $FEV_{1.0}$ is the difference between the volume at the breakpoint and the volume at that point on the curve which occurred 1 second into the forced exhalation. To find that point, measure the distance between the vertical lines on the paper that indicate 1-second time. Then measure the same distance horizontally from the breakpoint and draw a vertical line from there to where it crosses the exhalation tracing. Note the volume on the tracing where the two lines cross. In Figure 4.4, this is at a volume of 3,460 ml. The difference between this number and the volume at the breakpoint is 2,310 ml, which is the $FEV_{1.0}$ for this tracing.
14. Calculation of the $FEV_{1.0}\%$. Simply divide the $FEV_{1.0}$ by the FVC and multiply the fraction by 100.
15. Calculation of FEF_{25-75}. The FEF_{25-75} is simply the average flow rate during the middle 50% of the forced expiratory spirogram. To calculate this value, you need to know the slope of the line connecting the points on the expiratory tracing between 25% and 75% of the forced vital capacity. Determine the values associated with 25% and 75% of FVC by multiplying the FVC determined above by 0.25 and 0.75, respectively. Then add these numbers to the volume at the breakpoint

in the expiratory tracing to find the volumes on the tracing that correspond to 25% and 75% of exhalation. On the tracing in Figure 4.4, these are at 1,895 and 3,385 ml. When you have found these points, draw a line that connects those two points and also crosses two adjacent time lines. The difference between the points where your straight line crosses the 1-second lines is equal to the slope of your line in ml/sec. In Figure 4.4 these points are 2,510 and 4,580 ml, and the difference is 3,385 ml/sec.

16. When you have calculated each of the values on the three best tracings, enter these numbers in the appropriate blanks on your worksheet. The FVC reported for your subject is the single best effort. The $FEV_{1.0}$ reported is the single best effort, as well. This will often be from the same tracing as the best FVC, but not always. The $FEV_{1.0}$% value is obtained from the best FVC and the best $FEV_{1.0}$, even if they are not from the same curve. The FEF_{25-75} reported is the one from the spirogram with the best combination of FVC and $FEV_{1.0}$.

MAXIMAL VOLUNTARY VENTILATION (MVV)

During this maneuver, the subject breathes maximally from room air into a meteorological balloon. The duration of the effort is usually 12 seconds. Following are the steps to perform the MVV test:

1. The assembly of the apparatus to collect MVV is relatively simple. Connect a length of tubing at least 1.25″ in diameter to the expiratory port of a two-way breathing valve. Then connect a three-way directional stopcock valve to the tubing, and finally a meteorological balloon to an open port of the stopcock.
2. Have the subject seated comfortably with the apparatus. Instruct the subject to breathe normally through the apparatus for several seconds, during which time the stopcock should be open to room air. Then have the subject begin breathing as deeply and rapidly as possible. After the end of the first exhalation turn the stopcock valve to direct air into the balloon.
3. At the end of the 12-second time period, turn the stopcock back so the subject breathes into the room and tell the subject to relax.
4. Measure the volume of air in the balloon with a volume meter or Tissot spirometer.
5. Convert the volumes that you have measured to L/min by multiplying by 5 since MVV was measured for 12 seconds (1/5th of a minute). Record this value on your worksheet.

This maneuver takes some practice on the part of both the subject and the investigator. As with the forced spirometry tests, coaching of the subject is usually necessary. If the subject becomes dizzy, stop the test and allow recovery. Common mistakes include not giving the subject sufficient rest between trials, turning the stopcock while the subject is in mid-exhalation, and lack of effort on the part of the subject. The test should be performed at least three times.

MEASUREMENT OF RESIDUAL VOLUME

There are a number of methods to measure residual volume (RV) or functional residual capacity, but most require specialized equipment not found in the typical exercise physiology laboratory. The method described here (Wilmore et al., 1980) has the subject breathe in and out of a rubber or vinyl bag several times and measures the equilibration of N_2 between the lungs and the bag. It incorporates supplies and equipment found in many laboratories and is therefore applicable to many testing situations. In order to measure residual volume:

1. The rebreathing apparatus used to measure RV consists of a three-way stopcock breathing valve with a mouthpiece. Attached to the valve is a small (5L) meteorological balloon or rebreathing bag. Have this assembled prior to the subject's arrival.
2. Flush the rebreathing bag at least three times with 100% oxygen, then fill it with 3–5 liters of oxygen. The filling volume should be approximately 80% of the subject's vital capacity. The actual volume that you put into the rebreathing bag is not critical, but it is critical that you measure that volume carefully and note it in your worksheet.
3. Have the subject seated comfortably with nose clip and mouthpiece in place.
4. Instruct the subject to breathe normally 3–5 times, then exhale to residual volume.
5. When the subject indicates attainment of residual volume by raising the index finger, turn the stopcock so that the subject is connected to the rebreathing bag.
6. Have the subject take 5–7 deep breaths at a rate of 2 sec per breath, then exhale to residual volume at the end of the last breath.
7. When the subject reaches residual volume, switch the valve so that the subject is back to breathing room air.
8. Measure the O_2 and CO_2 concentrations of the rebreathing bag and note them in the worksheet. Calibration of the O_2 and CO_2 analyzers is described in Appendix C. Calculate residual volume using the following equation:

$$RV = \frac{VO_2 \times b}{79.8 - b}$$

Where: VO_2 = the volume of oxygen in the rebreathing bag at the beginning of the measurement.

b = the percentage of nitrogen in the rebreathing bag after the measurement; calculated as $100\% - \% O_2 - \% CO_2$.

CONVERSION TO STANDARD UNITS (BTPS)

The standard measurement units for pulmonary volumes and flow rates are in BTPS, or Body Temperature and Pressure Saturated, which is ambient or body barometric pressure, 37°C, and 100% saturated with water vapor. This standard is used to allow for comparison between laboratories at conditions under which the body exists, and all pulmonary function measurements (flows or volumes) are reported in BTPS units. When you measure the various volumes we have discussed, they are under ambient temperature, pressure and saturation (ATPS) conditions, which vary from day-to-day and between laboratories. The way to convert ATPS volumes to BTPS is with the following equation:

$$V_{BTPS} = V_{ATPS} \times BTPS_{CF}$$

Where:

V_{BTPS} is volume in BTPS units

V_{ATPS} is volume in ATPS units, volume measured in spirometer or volume meter

$BTPS_{CF}$ is the BTPS corrections factor, calculated below:

Where:

$$BTPS_{CF} = \frac{BT(°C)+273}{RT(°C)+273} \times \frac{P_B - (P_{H_2O})_{RT}}{P_B - (P_{H_2O})_{BT}}$$

BT = Body temperature in degrees Celsius (assume 37°C)
RT = Room temperature in degrees Celsius
273 = Conversion from degrees Celsius to degrees Kelvin
P_B = Barometric pressure
P_{H_2O} = Water vapor pressure at body temperature (BT) or room temperature (RT). This can be obtained from Table 4.1.

TABLE 4.1 — Water vapor pressures of saturated air at various temperatures

TEMPERATURE (°C)	WATER VAPOR PRESSURE (mmHg)
18	15.6
20	17.5
22	19.8
24	22.4
26	25.2
28	28.3
30	31.8
37	47.1

These numbers should be entered on the worksheet to follow.

COMPARISON TO NORMAL VALUES

Now that you have all of your measurements of pulmonary function, you still need to know what they say about the pulmonary health of the subject you have just tested. This can be done by comparing the values that you obtained with those shown by previous work to be within the normal range for a healthy population. There are a number of standard equations with which to compare the values obtained. All have been carefully collected from large populations of subjects of both sexes and of varying ages and sizes; all have a certain amount of error potential because of the variability in populations. Currently, the values accepted by the Occupational Safety and Health Administration (OSHA) are those of Knudsen et al. (1976), who provided prediction equations for FVC, $FEV_{1.0}$, and $FVC/FEV_{1.0}$ ($FEV_{1.0}\%$). These are given in Table 4.2.

TABLE 4.2 — Calculations for predicted forced spirometric values (from Knudsen et al., 1976), and maximal voluntary ventilation (from Taylor et al., 1989)

FEMALES UNDER 20 YEARS	MALES UNDER 25 YEARS
FVC = 0.033.H + 0.092.A − 3.469	FVC = 0.050.H + 0.078.A − 5.508
$FEV_{1.0}$ = 0.027.H + 0.085.A − 2.703	$FEV_{1.0}$ = 0.046.H + 0.045.A − 4.808
$FEV_{1.0}$% = 107.38 − 0.111.H − 0.109.A	$FEV_{1.0}$% = 103.64 − 0.087.H − 0.140.A
FEMALES 20 YRS AND OVER	**MALES 25 YRS AND OVER**
FVC = 0.037.H − 0.022.A − 1.774	FVC = 0.065.H − 0.029.A − 5.459
$FEV_{1.0}$ = 0.027.H − 0.021.A − 0.794	$FEV_{1.0}$ = 0.052.H − 0.027.A − 4.203
$FEV_{1.0}$% = 107.38 − 0.111.H − 0.109.A	$FEV_{1.0}$% = 103.64 − 0.087.H − 0.140.A
MVV = 0.55.H − 0.72.A + 50	MVV = 1.15.H − 1.27.A + 14

Where: H = height in centimeters.
        A = age in years.

QUESTIONS AND ACTIVITIES

1. Make a table of the $FEV_{1.0}$, FVC, and MVV for all class members.
2. Graph $FEV_{1.0}$ versus MVV for the class. Is there a relationship between the two measurements? Would you expect this? Why or why not?
3. Graph FVC versus height for males and females separately. Is there a relationship? Is this relationship different for males and females? How do you know if they are different or not?
4. The measurement of residual volume depends on the dilution of a gas in the lungs. What would happen to the measurement if portions of the airways were blocked, as in certain pulmonary disease?
5. We use forced spirometry values for clinical assessment of subjects rather than for assessments of athleticism. Why would these measurements not be good predictors of athletic ability?
6. Why are nose clips necessary during pulmonary function measurements?
7. If a subject had mild asthma with some narrowing of the airways, which of the forced spirometry measurements would likely be affected?

REFERENCES

Ferris, B. "Epidemiology Standardization Project." *American Review of Respiratory Disease.* 118 (1978): 1–120.

Knudsen, R. J., R. C. Slatin, M. D. Lebowitz, and B. Burrows. "The Maximal Expiratory Flow-Volume Curve: Normal Standards, Variability and Effects of Age." *Am Rev Respir Dis* 113 (1976): 587–600.

Schwartzstein, R. M., and M. J. Parker. *Respiratory Physiology, a Clinical Approach.* Philadelphia: Lippincott, Williams and Wilkins, 2006.

Wilmore, J. H., P. A. Vodak, R. B. Parr, R. N. Girandola, and J. E. Billing. "Further Simplification of a Method for Determination of Residual Lung Volume." *Med Sci Sports Exerc* 12 (1980): 216–18.

PULMONARY FUNCTION WORKSHEET

Forced Volumes (ATPS)

	Vol at Brkpt (A)	Vol at plateau (B)	Vol at 1 sec (C)	Vol at 25% VC (D)	Vol at 75% VC (E)	Slope of Line (S) (D–E)/sec	FVC (ml) (B–A)	FEV$_{1.0}$ (ml) (C–A)	FEF$_{25-75}$ (ml/sec) (S)
Trial 1	___	___	___	___	___	___	___	___	___
Trial 2	___	___	___	___	___	___	___	___	___
Trial 3	___	___	___	___	___	___	___	___	___
Trial 4	___	___	___	___	___	___	___	___	___

Maximal Voluntary Ventilation (ATPS)

	Bag Volume (L)	×	4 5	= MVV (L/min ATPS)
Trial 1	44.3	×	5	= 177.2 (L/min ATPS) 191.02 L/min BTPS
Trial 2	_____	×	5	= _____ (L/min ATPS)
Trial 3	_____	×	5	= _____ (L/min ATPS)

Residual Volume Calculation

Nitrogen in bag at equilibration

	100%	%O$_2$	%CO$_2$	= % Nitrogen
Trial 1	100%	___	___	= ___
Trial 2	100%	___	___	= ___
Trial 3	100%	___	___	= ___
Trial 4	100%	___	___	= ___

Height is on x-axis
2 and 3 have scatter plots and use trend line

FVC = Actual/Predicted > 80%, no RLD

FEV$_{1.0}$ = Actual/Predicted > 80%, no OLD

FEV$_{1.0}$/FVC > 70%, no OLD

saturated air - water in air

Room Temp = 23°C = T$_R$
P$_B$ = 595.2 mmHg

FVC = L ATPS BTPS
FEV$_{1.0}$ = L ATPS BTPS
MVV = L/min ATPS BTPS

1.078 BTPS

Residual Volume

	Bag Volume (L)	×	%N$_2$	÷	(79.8 − % N$_2$)	=	RV	(L, ATPS)
Trial 1	_____	×	_____	÷	(79.8 − ____)	=	_____	(L, ATPS)
Trial 2	_____	×	_____	÷	(79.8 − ____)	=	_____	(L, ATPS)
Trial 3	_____	×	_____	÷	(79.8 − ____)	=	_____	(L, ATPS)
Trial 4	_____	×	_____	÷	(79.8 − ____)	=	_____	(L, ATPS)

BTPS Correction Factor (BTPS$_{CF}$)

$310 \div (\text{Room Temp} + 273) \times [P_B - (P_{H_2O})_{RT}] \div [P_B - (P_{H_2O})_{BT}] = \text{BTPS}_{CF}$

$310 \div (\underline{} + 273) \times [\underline{} - \underline{}] \div [\underline{} - \underline{}] = \underline{}$

PULMONARY FUNCTION SUBJECT DATA SHEET

Name: _____

Date: _____

Age: _____   Height (cm): _____

Sex: _____   Smoker? _____

Room temperature (°C): _____   Barometric pressure (mmHg): _____

Technician: _____

	Actual	Predicted	% Predicted
FVC (L, BTPS)	_____	_____	_____
$FEV_{1.0}$ (L, BTPS)	_____	_____	_____
$FEV_{1.0}\%$	_____	_____	_____
FEF_{25-75} (L/sec, BTPS)	_____	_____	_____
MVV (L/min, BTPS)	_____	_____	_____
Residual Vol (L, BTPS)	_____		

CHAPTER 5

THE MEASUREMENT OF OXYGEN UPTAKE AND ENERGY EXPENDITURE

OUTCOME OBJECTIVES

Following successful completion of this exercise, the student should be able to:

1. Understand the meaning and derivation of the oxygen uptake measurement and the measurement of energy expenditure.
2. Perform the techniques necessary to measure oxygen uptake, carbon dioxide production, and respiratory exchange ratio.
3. Use the values of oxygen uptake and carbon dioxide production to calculate energy expenditure during a task.

INTRODUCTION

For tasks lasting longer than just a few minutes, the primary source of energy is through aerobic metabolism. The amount of ATP produced in the muscles and other tissues is directly related to the oxygen that is used in the last step of oxidative phosphorylation and the CO_2 produced from the metabolism of organic compounds in the Krebs cycle. Since we do not store large quantities of oxygen within the body, we must deliver the oxygen to the tissues as it is being used in order to maintain an adequate prolonged energy supply. Therefore the oxygen uptake ($\dot{V}O_2$) during exertion is probably one of the most widespread measurements made in the exercise physiology laboratory.

The uses of oxygen uptake measurements are also widespread. The principal use of $\dot{V}O_2$ is to calculate energy expenditure. These measurements may be made at rest or during periods of work or exercise. When $\dot{V}O_2$ is employed for measuring energy expenditure this is referred to as "indirect" calorimetry, which is different from "direct" calorimetry, the measurement of the production of heat by the body. When the $\dot{V}O_2$ and energy expenditure are calculated, they can be used in athletics, in the workplace, and in hospitals to determine the capability of the body to perform work, to measure the efficiency of performing various activities (efficiency = work done ÷ energy used) and to assess certain metabolic abnormalities associated with some disease states. The maximal oxygen uptake ($\dot{V}O_2$max) is important in that it is the primary indicator of the aerobic fitness of a subject.

As with any measurement, however, there are a number of assumptions to make in order for $\dot{V}O_2$ to mean what it is intended to mean. While $\dot{V}O_2$ measurements can be made at any time, they have to be made during a "steady-state" period for them to be valid indicators of the energy expenditure during a task. This means that whatever is being done while you measure $\dot{V}O_2$ must be done for about 3 minutes or longer. If the duration of the exercise is less than that, the $\dot{V}O_2$ will often still be rising and the measurement you make will be lower than the actual oxygen uptake needed for that task. We also assume that fats and carbohydrates make up the entire fuel source for the body and that the contribution of protein as a fuel is negligible. In most cases, protein makes up a very small portion of the total fuel used, and its measurement is beyond the scope of this text. The error that would be made by not measuring its contribution is quite small, though, and thus is ignored. In some instances, however, protein may make up a substantial portion of the fuel for exercise, such as starvation (dieting) or cases of very high protein intake, and adjustments must be made when using the $\dot{V}O_2$ to calculate energy usage. Another limitation, as with any measurement you will make in the laboratory, is the equipment. Your measurement will not be correct if you do not calibrate the equipment you use to measure $\dot{V}O_2$.

Most labs use automated metabolic carts of various types to measure $\dot{V}O_2$ and $\dot{V}CO_2$. The use and calibration of these is as varied as the equipment. All consist of calibrating the equipment used to measure breathing volume as well as the O_2 and CO_2 concentrations of the expired gases. In this section we will describe the most basic system, collection in bags.

EQUIPMENT USED AND CALIBRATION

Incumbent on the measurement of many metabolic variables is obtaining the correct analysis of gas samples. As you will see, getting accurate measurements of oxygen consumption and carbon dioxide production ($\dot{V}CO_2$) demand precise gas analysis, with respect to both content and volume. The major pieces of equipment needed to measure $\dot{V}O_2$ and $\dot{V}CO_2$ are analyzers for oxygen and carbon dioxide and an apparatus to measure gas volumes.

MEASUREMENT OF GAS FRACTIONS

The measurement of concentration of inspired and expired gases has historically been done in a number of ways. Among the earliest was the weighing of samples. If a volume of gas is weighed, then one gas is removed (oxygen for example), the weight of the remaining gas will be less. This difference, divided by the original weight, reflects the concentration of oxygen in the sample. This method had its problems, though, in that large samples and very accurate scales were required.

By the same token, if one gas in a mixture is removed, the volume will change. This change will also reflect the concentration of the gas removed. This is the method by which many manual gas analyzers work. One such analyzer is called a Scholander apparatus. The method is simple, although its operation is relatively tedious. Basically, the Scholander apparatus consists of a reaction chamber, to which the gas is introduced, and chambers containing chemicals that absorb oxygen or carbon dioxide. If one of the gases is absorbed by the chemical, the volume of the total gas is decreased, and that decrease is measured by the apparatus. This is described in more detail in Scholander's original article (Scholander, 1947): "Analyzer for accurate estimation of respiratory gases in one-half cubic centimeter samples." For truly accurate calibration of your gas analyzers, you should have calibration gases that have been standardized by an apparatus such as the Scholander, since the authors have purchased calibration gases from some suppliers that had actual gas concentrations different from the stated levels by as much as 0.5%–1%.

The method you will use to measure gas fractions is by electronic analysis. These analyzers, one for oxygen and one for carbon dioxide, depend on calibration with known gases in order to give correct results: hence the importance of an apparatus such as the Scholander, even in the most automated of laboratories. Operation and calibration of these analyzers is described in Appendix C.

MEASUREMENT OF GAS VOLUMES

Just as important as the correct analysis of gas fraction is the accurate measurement of the volume of a gas sample. Methods for doing this range from those that are extremely simple, to those which are complex and require numerous electronic manipulations.

The simplest method of measuring gas volumes is with the Tissot spirometer. This device is similar to, but larger than, the spirometers used in clinical spirometry. It consists of a bell inverted in a cylinder filled with water. As air is put into the Tissot, the bell rises, and this displacement is measured by the meter stick on the side. A tank conversion factor is stamped onto the Tissot to allow conversion of millimeters displacement of the bell into liters of volume put into the Tissot. While some labs still use this technique, most Tissot spirometers gather dust and rust in storage units.

The next method for gas volume measurement is by the use of a mechanical volume meter. This is typified by the Parkinson-Cowan dry gas meter, which is similar to meters you have seen that measure the natural gas used in a home. These meters measure the volume of air forced through them and convert the flow to a sweep of the dial. Each full revolution equals 10 liters. These meters must periodically be calibrated against a calibration syringe, typically 3L in volume. To use this meter, record the beginning reading from the dial meters. Then empty the sample bag through the inlet port on the meter. Record the final dial reading, and subtract the initial from the final readings to get volume.

If you have neither a Tissot spirometer nor a volume meter and want to use the bag technique, you can evacuate the bags with a 3L syringe. These are also very accurate, though this technique is more time consuming.

CONVERSION FROM AMBIENT TO STANDARD CONDITIONS

Because of the fact that ambient temperature, humidity, and barometric pressure vary, there is a need for conversion of these measurements to some sort of standard conditions. This conversion allows for two things. One is the comparison of values obtained in different places and under different ambient conditions. The other is for the conversion of oxygen consumption to caloric expenditure. Since the equations for burning carbohydrates or fats have all been converted to standard conditions, the measurement of oxygen consumption and carbon dioxide production must be as well.

The standard condition used for oxygen uptake measurements is referred to as **STPD** or **S**tandard **T**emperature and **P**ressure, **D**ry. This is used for all metabolic volumes, such as $\dot{V}O_2$ or $\dot{V}CO_2$. STPD conditions are 0° C, 0% humidity, and 760 mmHg barometric pressure. Measurements are made under **ATPS** conditions or **A**mbient **T**emperature and **P**ressure **S**aturated. The conversion from ATPS to STPD is done by the following equation:

$$\text{Volume STPD} = \text{Volume ATPS} \times \frac{P_b - P_{H_2O}}{760} \times \frac{273}{T_a (\deg K)}$$

Where:

P_b = barometric pressure
P_{H_2O} = water vapor pressure under current temperature and humidity conditions
T_a = ambient temperature in degrees Kelvin (deg C + 273)

TABLE 5.1 – Relationship between temperature and water vapor pressure

Temperature		P_{H_2O} Saturated
(deg C)	(deg K)	(mmHg)
20	293	17.5
22	295	19.8
24	297	22.4
26	299	25.2
28	301	28.3
30	303	31.8
32	305	35.6
34	307	39.9
36	309	44.6
38	311	49.7

P_{H_2O} under saturated conditions is determined by the ambient temperature and is given in Table 5.1. To get ambient P_{H_2O}, multiply this number by the relative humidity.

With the knowledge of the equipment used and how to calibrate that equipment, we can now proceed to the calculations for the measurement of oxygen uptake and carbon dioxide production. If these are known and understood, it is much easier to understand how the apparatus for collecting and measuring $\dot{V}O_2$ and $\dot{V}CO_2$ is set up.

CALCULATIONS

1. OXYGEN CONSUMPTION

Oxygen consumption is the difference between the volume of oxygen inhaled and that exhaled, that is, the amount used. Therefore, in its strictest sense:

$$\dot{V}O_2 = O_2 \text{ in} - O_2 \text{ out}$$

or more reasonably:

$$\dot{V}O_2 = (\dot{V}_I \times F_I O_2) - (\dot{V}_E \times F_E O_2)$$

Where:

$\dot{V}O_2$ = oxygen consumption (L/min or ml/min)
$\dot{V}_I$ = volume of air inhaled (L/min or ml/min)
$F_I O_2$ = fraction of O_2 in inhaled air (typically 0.2093)
$\dot{V}_E$ = volume of air exhaled (L/min or ml/min)
$F_E O_2$ = fraction of O_2 in exhaled air

This is a simple procedure except for the simultaneous measurement of $\dot{V}_I$ and $\dot{V}_E$. If you measure one or the other and assume the two to be equal, you will make mistakes, because under some circumstances inspired and expired ventilations are quite different. It is relatively simple to know both $\dot{V}_I$ and $\dot{V}_E$ while only measuring one. This is based on the nitrogen concentrations (F_IN_2 and F_EN_2). Since N_2 is neither used or given off:

$$N_2 \text{ in} = N_2 \text{ out}$$

$$\text{or: } \dot{V}_I \times F_IN_2 = \dot{V}_E \times F_EN_2$$

Where:

$$F_IN_2 = \text{fraction of } N_2 \text{ in the inspired air, or } F_IN_2 = 1 - F_IO_2 - F_ICO_2$$
$$F_EN_2 = \text{fraction of } N_2 \text{ in the expired air}$$

This is called the Haldane transformation and is quite commonly used. By rearranging the equations, we can get:

$$\dot{V}_I = \dot{V}_E \times \frac{F_EN_2}{F_IN_2} \text{ or } \dot{V}_E = \dot{V}_I \times \frac{F_IN_2}{F_EN_2}$$

This can now be substituted into the earlier equation as follows:

$$\dot{V}O_2 = (\dot{V}_I \times F_IO_2) - \frac{(\dot{V}_I \times F_IN_2)}{F_EN_2} \times F_EO_2$$

or

$$\dot{V}O_2 = \frac{(\dot{V}_E \times F_EN_2)}{F_IN_2} \times F_IO_2 - (\dot{V}_E \times F_EO_2)$$

2. CARBON DIOXIDE PRODUCTION

The equation for $\dot{V}CO_2$ is similar to that for $\dot{V}O_2$. Basically:

$$\dot{V}CO_2 = CO_2 \text{ out} - CO_2 \text{ in}$$

or

$$\dot{V}CO_2 = (\dot{V}_E \times F_ECO_2) - (\dot{V}_I \times F_ICO_2)$$

The Haldane transformation is used here as well as in the calculation of $\dot{V}O_2$. $\dot{V}CO_2$ is an important measure for helping in determining the foodstuff being used for fuel, as will be seen below, and for the conversion of $\dot{V}O_2$ to energy expenditure.

3. RESPIRATORY EXCHANGE RATIO

The respiratory exchange ratio (RER) is the ratio of the $\dot{V}CO_2$ to the $\dot{V}O_2$. Because of the chemical makeup of fats and carbohydrates, different amounts of oxygen are used to generate the same amount of carbon dioxide when the fuels are broken down. In addition, different amounts of energy are produced by the breakdown of fats versus carbohydrates. Therefore, RER is used to determine what food is being used as a fuel by the subject and how many kilocalories of energy a subject is producing. An RER of

0.7 indicates that the subject is using 100% lipids for energy. An RER of 1.0 indicates that the subject is using all carbohydrates as fuel. This ratio assumes that little or no protein is used. This is the reason for the assumption mentioned in the earlier part of the exercise. The equation for RER is given below, as well as the fuels used and the energy produced by using 1 liter of oxygen (see Table 5.2).

$$RER = \frac{\dot{V}CO_2}{\dot{V}O_2}$$

TABLE 5.2 – Fuel composition and caloric equivalent of one liter of oxygen at various RER values

RER	Caloric Equivalent of Oxygen		% Carbohydrate as a Fuel	% Fat as a Fuel
	(kcal/liter of O_2)	(kJ/liter of O_2)		
0.70	4.686	19.62	0	100.0
0.72	4.702	19.68	4.8	95.2
0.74	4.727	19.79	12.0	88.0
0.76	4.751	19.89	19.2	80.8
0.78	4.776	19.99	26.3	73.7
0.80	4.801	20.10	33.4	66.6
0.82	4.825	20.20	40.3	59.7
0.84	4.850	20.30	47.2	52.8
0.86	4.875	20.41	54.1	45.9
0.88	4.899	20.51	60.8	39.2
0.90	4.924	20.60	67.5	32.5
0.92	4.948	20.71	74.1	25.9
0.94	4.973	20.82	80.7	19.3
0.96	4.998	20.92	87.2	12.8
0.98	5.022	21.02	93.6	6.4
1.00	5.047	21.13	100.0	0

Based on data from N. Zuntz and H. Schumberg. In G. Lusk, *The Elements of the Science of Nutrition*, 4th ed. Philadelphia: W.B. Saunders, 1928, p. 65.

PROCEDURES FOR MEASURING $\dot{V}O_2$

From the equations given, you can see that in order to calculate the amount of oxygen consumed or CO_2 produced by a subject, we need to measure certain variables. We will have to measure the amount of air either inhaled or exhaled, and the oxygen and CO_2 fractions in both the inhaled and exhaled air. A number of methods have been used to do this, ranging from very simple collections of air into bags, to computer-operated methods that make all the necessary calculations as often as every breath. This exercise will use one of the simplest techniques available, the Douglas bag open circuit technique. We will also briefly cover some of the other methods used. To measure $\dot{V}O_2$ and $\dot{V}CO_2$ by the Douglas bag technique, perform the following steps:

WARNING

Do not do strenuous exercise tests on any subject over 45 years of age or with any history of cardiovascular or respiratory disease or high blood pressure. Cease testing if the subject feels faint or nauseous. Inquire as to the subject's perceptions of the workload with every load and discontinue testing if the subject does not feel he or she can or wants to continue.

1. If you are going to measure exercise $\dot{V}O_2$, have the exercise apparatus (cycle ergometer or treadmill) set up prior to beginning the measurements. Then, set up the open circuit apparatus as shown in Figure 5.1. This consists of a two-way breathing valve with a mouthpiece attached and tubing attached to the expiratory port. At the end of the expiratory tubing, attach a two-way breathing stopcock so that when the subject exhales, air will exit one or the other port, depending on which way the valve is turned. To the closed port of the stopcock valve, attach a Douglas bag or meteorological balloon that has had all of the air evacuated from it. Some care must be taken with the attachment of both the breathing valve and the stopcock valve to make sure they are placed so that the subject can inhale from room air and exhale into the bag for gas collection.
2. Make sure that the analyzers for gas concentration and volume are calibrated using standard gases and a known volume calibration syringe. This should be done within 10–20 minutes of sampling, since many analyzers have some "drift" and will go out of calibration periodically.
3. Measure and record the room temperature and barometric pressure.
4. Place the mouthpiece in the subject's mouth. Make sure that the subject also wears nose clips while expired gases are being collected.
5. Have the subject begin exercising. It is best to provide some warm-up for the subject unless the early portions of the exercise bout consist of only light exercise. To ensure that the exercise is steady state, you should collect gas after approximately 3–5 minutes at a load.

FIGURE 5.1 — Diagram of system for measuring oxygen uptake by collecting expired gases in a meteorological balloon.

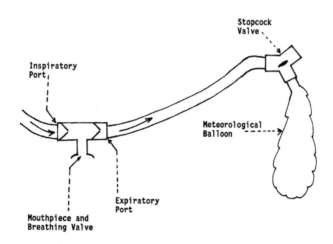

6. To begin the gas collection, turn the stopcock valve so that the subject can exhale into the collection bag. When you turn this valve, note the time to the second, or start a timer so that you will know the exact amount of time for which you have collected the gas.
7. At the end of the collection period, turn the stopcock back to the position in which the port to the collection bag is closed. Again, note the time at which this is done. The timing of the collection is not critical, although a standard collection time is 1 minute, since ventilation and $\dot{V}O_2$ values are generally expressed in L/min. It is critical, though, that the exact time is known. Record the collection time in the appropriate space on your worksheet.
8. Remove the bag gently from the valve and seal it with a stopper to prevent leakage. If you are going to collect other samples, the bag can be safely stored for up to approximately 30 minutes without a change in gas concentrations.
9. To analyze the contents of the expired collection bag, attach the sampling tube of the previously calibrated gas analyzer to the bag via a stopcock attached to the neck of the bag or through the stopper. Switch the stopcock to sample from the collection bag. Be sure to time the duration that the gas analyzer is attached to the collection bag. This is important because gas analyzers draw gas from the bag, and this will decrease the volume that you measure. Record this time on your worksheet. Most gas analyzers sample at a rate of 100–500 ml/min. Your operating manual will tell you the sample rate.
10. When the readout on the gas analyzers stabilize, record the expired O_2 and CO_2 concentrations in the spaces on the worksheet. Make sure that you convert the values to a decimal fraction rather than using the percent values given on most gas analyzer readouts.
11. Record the values obtained for the O_2 and CO_2 concentrations of room air, in the appropriate spaces on your worksheet. These values are typically 0.2093 for O_2 and 0.0003 for CO_2.
12. Next, measure the volume of the collected gas. If you use a Tissot spirometer record the displacement of the meter stick and multiply that value by the tank calibration factor. If you use a dry gas meter, just read the volume displacement from the dials. This will be the ATPS volume of your bag. Record this value in the appropriate spot on your worksheet.
13. Add in the volume taken by the gas analyzers. This is the product of the analyzer flow rate and the sampling time. The sum gives you the expired volume in liters (ATPS). Record this value on your worksheet. If the gas collection time was different from 1 minute, convert to liters per minute by dividing the volume by the collection time in seconds and multiplying by 60:

$$\text{Volume (L/min)} = \text{Bag volume (L)} \div \text{collection time (sec)} \times 60 \text{ sec/min}$$

14. Convert the minute ventilation in ATPS to STPD using the equation provided previously. Remember, if you have collected exhaled air to measure volume, the P_{H_2O} is the value given in the table because exhaled air is 100% saturated with water. Record this value on your worksheet.
15. Calculate $\dot{V}O_2$, $\dot{V}CO_2$, and RER according to the equations presented. Record these values on your worksheet.
16. To calculate the energy expenditure, multiply the value you obtained for $\dot{V}O_2$ by the caloric equivalent of oxygen (kcal/liter O_2) found in Table 5.2 at the RER calculated. Record this value as well.

Make the measurements at several loads or grades on the exercise apparatus.

SOURCES OF ERROR

Although the method of measuring $\dot{V}O_2$, $\dot{V}CO_2$ and RER for this experiment (the Douglas bag technique) is one of the simplest to perform, errors can still be made in the measurements. Many possible errors have been addressed with the steps presented in the exercise. One source of error that cannot be overemphasized is the lack of proper calibration of equipment. Methods for calibrating gas analyzers and volume monitors are given in this text and should be followed with each use. Failure to properly calibrate equipment effectively invalidates any measurements obtained from that equipment. Other sources of error include improper timing of either collection or sampling times, which then leads to systematic increases or decreases in the volumes, and leaks in the equipment, either at the bag or in the tubing. All of these variables need to be checked when making measurements that demand accuracy.

OTHER $\dot{V}O_2$ MEASUREMENT DEVICES

The method described in this exercise was one of the simplest ways possible to measure oxygen uptake. A number of other methods exist to measure $\dot{V}O_2$ as well. A very common method employs a mixing chamber on the exhaled side, from which exhaled gases are sampled continuously. In this method, volume is measured on the inspired side, and the combination of P_{H_2O} from the table above and relative humidity are used to determine the STPD correction. A diagram of this setup is shown in Figure 5.2. This method also lends itself to computerization, because the continuous gas sampling and continuous monitoring of inspired volume are easily fed into a computer for calculation of $\dot{V}O_2$ and $\dot{V}CO_2$. Different manufacturers have different calibration procedures. Make sure you follow those procedures exactly.

Another relatively common method of measuring $\dot{V}O_2$ is dependent on computerization to a great extent, and many of the newer metabolic measurement carts employ variations of this method. Basically, inspired and/or expired breath volumes are measured using flow or volume monitors near the mouthpiece, and gases are sampled near the mouthpiece as well. These values are then fed into a computer and all calculations are processed at regular intervals, or even on a breath-by-breath basis. These systems have the advantage of providing a large amount of information rapidly and may be quite useful, particularly in clinical situations, where many subjects are evaluated in a relatively short period of time. They may also measure some variables that would take large amounts of time if done by hand. The cost of these systems can sometimes put them out of the range of many laboratories, and the calibration and

FIGURE 5.2 — Diagram of mixing chamber setup for the measurement of oxygen uptake.

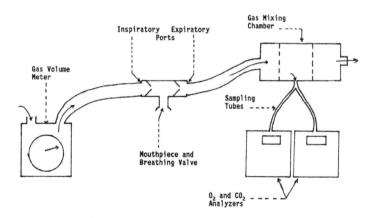

FIGURE 5.3 — Fully automated oxygen consumption system.

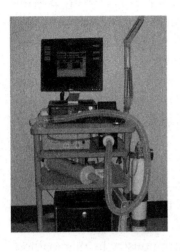

troubleshooting can be a disadvantage in these complicated systems. A typical computerized metabolic measurement setup is shown in Figure 5.3.

QUESTIONS

1. What is meant by "steady-state" and why is the attainment of a steady-state important in measuring $\dot{V}O_2$?
2. What does the RER tell us? What does it mean if RER is greater than 1.0? What happens to RER with increased exercise intensity? What would you expect to happen to RER if you had consumed a bowl of pasta an hour before the measurement? How about several slices of cheese?
3. How are $\dot{V}O_2$ and $\dot{V}CO_2$ related?
4. Is there a relationship between $\dot{V}O_2$ and power output? Why?

REFERENCES

Scholander, P. "Analyzer for Accurate Estimation of Respiratory Gases in One-Half Cubic Centimeter Samples." *J. Biol. Chem.* 167 (1947): 235–40.

$$RER = \frac{VCO_2}{VO_2} = \frac{1.08}{1.26} = 0.86$$

@ 0.86   4.875 kcal/L × 1.26 L/min = kcal/min

VO_2 Max

1) Plateau of VO_2 or 2 of 3
2) RER ≥ 1.10
3) RPE ≥ 17
4) HR within 10 beats of age-predicted HR max
   - HR max = 220 - age

$$V_{STPD} = V_{ATPS} \times \frac{P_B - P_{H_2O}}{760} \times \frac{273}{(273 + RT)}$$

$P_B = 596.7$ mmHg
$RT = 26°C$

$$= 25.74 \times \left(\frac{596.7 \text{ mmHg} - 25.2}{760}\right) \times \frac{273}{299}$$

$= 25.74 \times 0.752 \times 0.913$

$V_E = 17.67$ L/min STPD

$$\frac{V_I \times F_I N_2 = V_E \times F_E N_2}{F_I N_2} \quad \frac{}{F_I N_2}$$

$F_I O_2 = .2094$ $F_E O_2 = 0.1404$
$F_I CO_2 = 0.0003$ $F_E CO_2 = 0.0598$
$F_I N_2 = 0.7903$ $F_E N_2 = 0.7998$

   in   -   out
$\dot{V}O_2 = (V_I \times F_I O_2) - (V_E \times F_E O_2)$

$= \left(\left(\frac{V_E \times F_E N_2}{F_I N_2}\right) \times F_I O_2\right) - (V_E \times F_E O_2)$

$= \left(\frac{17.67 \times .7998}{0.7903}\right) \times 0.2094 - (17.67 \times 0.1404)$

$= 1.26$ L/min STPD

MEASUREMENT OF $\dot{V}O_2$, $\dot{V}CO_2$, AND ENERGY EXPENDITURE WORKSHEET

Date: _____ Subject: _____

Barometric Pressure (P_B, mmHg): _____ Room Temperature (°C): _____

Room Air O_2 (F_IO_2): _____ Room Air CO_2 (F_ICO_2): _____

Room Air N_2 (F_IN_2): _____
($= 1 - F_IO_2 - F_ICO_2$)

Bag #	Collection Time (sec)	Bag O_2 (F_EO_2)	Bag CO_2 (F_ECO_2)	Bag N_2 (F_EN_2) ($1-F_EO_2-F_ECO_2$)	Bag Volume (L)
___	___	___	___	___	___
___	___	___	___	___	___
___	___	___	___	___	___
___	___	___	___	___	___

Minute Ventilation ($\dot{V}_E$)

$\dot{V}_E$ (L/min, ATPS) = [Bag Volume (L) ÷ Collection Time (sec) × 60 sec/min] + Analyzer Sample Volume (L)

Bag 1 $\dot{V}_E$: _____ Bag 2 $\dot{V}_E$: _____ Bag 3 $\dot{V}_E$: _____ Bag 4 $\dot{V}_E$: _____

$$\dot{V}_E \text{ (L/min) STPD} = \dot{V}_E \text{ (L/min) ATPS} \times \frac{P_b - P_{H_2O}}{760} \times \frac{273}{T_a \text{ (deg K)}}$$

Bag 1 $\dot{V}_E$ (STPD): _____ Bag 2 $\dot{V}_E$: _____ Bag 3 $\dot{V}_E$: _____ Bag 4 $\dot{V}_E$: _____

Oxygen Uptake ($\dot{V}O_2$)

$\dot{V}O_2$ (L/min, STPD) = [$\dot{V}_E$ (L/min, STPD) × F_EN_2 ÷ F_IN_2 × F_IO_2] − [$\dot{V}_E$ (L/min, STPD) × F_EO_2]

Bag 1 _____ = [_____ × ___ ÷ ___ × ___] − [_____ × ___]
Bag 2 _____ = [_____ × ___ ÷ ___ × ___] − [_____ × ___]
Bag 3 _____ = [_____ × ___ ÷ ___ × ___] − [_____ × ___]
Bag 4 _____ = [_____ × ___ ÷ ___ × ___] − [_____ × ___]

p.69

VCO_2 = out − in

$MET = \frac{\text{Relative } VO_2}{3.5 \, mL/kg/min}$

$\frac{L}{min} \times \frac{1000 \, mL}{L} \times \frac{1}{kg}$

$VO_2 = \dot{V}_E$

Carbon Dioxide Output ($\dot{V}CO_2$)

$\dot{V}CO_2$ (L/min, STPD) = [$\dot{V}_E$ (L/min, STPD) × F_ECO_2] − [$\dot{V}_E$ (L/min, STPD) × $F_EN_2 \div F_IN_2 \times F_ICO_2$]

Bag 1 _____ = [_____ × ____] − [_____ × ____ ÷ ____ × ____]
Bag 2 _____ = [_____ × ____] − [_____ × ____ ÷ ____ × ____]
Bag 3 _____ = [_____ × ____] − [_____ × ____ ÷ ____ × ____]
Bag 4 _____ = [_____ × ____] − [_____ × ____ ÷ ____ × ____]

Respiratory Exchange Ratio (RER)

RER = $\dot{V}CO_2 \div \dot{V}O_2$

Bag 1 RER:_____ Bag 2 RER:_____ Bag 3 RER:_____ Bag 4 RER:_____

Energy Expenditure (kcal/min)

Energy Expenditure (kcal/min) = $\dot{V}O_2$ (L/min, STPD) × Caloric Equivalent at RER
(kcal/L O_2, from Table 5.2)

Bag 1 Energy:_____ Bag 2 Energy:_____ Bag 3 Energy:_____ Bag 4 Energy:_____

Fick equation: $VO_2 = Q \times (a - vO_2 \text{ difference})$
(artery, vein)

RESTING AND EXERCISE ELECTROCARDIOGRAPHY

CHAPTER 6

OUTCOME OBJECTIVES

Following successful completion of this exercise, the student should be able to:

1. Demonstrate the proper skin preparation, electrode placement, and lead configurations for recording 12-lead electrocardiograms (ECG) during exercise.
2. Describe and illustrate the normal electrocardiographic waveforms recorded from the 12-lead ECG, and correlate these recorded waveforms with contractile events occurring in the heart.
3. Measure ECG waveforms and list normal voltage and duration values.
4. Describe the components of a model for ECG interpretation.
5. Identify ECG criteria for an abnormal graded exercise test (GXT).
6. Review the resting and submaximal exercise ECG for abnormal findings.

INTRODUCTION

It is generally standard procedure to monitor and record the electrocardiogram (ECG or sometimes called EKG) when conducting a diagnostic or functional graded exercise test (GXT), especially if a symptom-limited maximal GXT is employed. Diagnostic interpretation of the ECG waveforms requires a high level of expertise, and is usually the responsibility of a physician familiar with exercise electrocardiography. However, certain common and/or dangerous irregularities should be familiar to all exercise testing personnel. For example, the properly interpreted resting and exercise ECG can provide valuable information related to myocardial ischemia and infarction, atrial and ventricular arrhythmias, myocardial hypertrophy, diseases of the heart muscle and valves, and effects of cardiac drugs. Professionals working in athletic settings need also be familiar with electrocardiography, since well-trained athletes may exhibit abnormal ECG findings at rest and during exercise (Crouse et al., 2009; Thompson, Gordon, and Pescatello, 2010). When coupled with other variables measured during the GXT, such as blood pressure and ratings of perceived exertion, the resting and exercise ECG is an essential component of the exercise prescription process to ensure safe and effective exercise for healthy individuals. In addition, the ECG can be a diagnostic aid for the physician in detecting the presence of heart disease or in assessing the effectiveness of a specific therapeutic treatment for heart patients (Thaler, 2010; Thompson, Gordon, and Pescatello, 2010).

Fundamentally, the ECG is simply a graphic recording of the electrical activity generated by the heart during each heartbeat. Impulse formation and propagation through the specialized conductive and muscle tissue of the heart produces weak electric currents that spread through the entire body. By applying electrodes to prescribed anatomic locations on the body, then connecting these electrodes via electrical wires to a specially equipped computer system, the weak electric impulses can be digitized, amplified, processed, and filtered. The processed impulses can then be stored electronically on computer disks, displayed on a monitor, and recorded on paper as the familiar ECG waveforms commonly evaluated by clinicians.

The type of equipment used to collect and display ECG data varies considerably in standard features, automation, and expense. The least costly machines generally consist of a single channel recorder without a monitor and minimal circuitry to reduce unwanted electrical interference and allow for only manual control of ECG recordings. These single-channel units are rarely found in contemporary laboratory or clinical settings. Modern ECG machines usually consist of a three-channel recorder, a display monitor, and a computer to analyze and process the electronic ECG signal. The computer can usually be programmed to control a motorized treadmill or stationary cycle ergometer, and to automatically analyze the ECG signal for heart rate or waveform abnormalities that might be clinically relevant. These computerized machines also provide for many manual or automated inputs at rest and during exercise, such as subject demographics, medications, medical history, blood pressure, blood-oxygen saturation, ratings of perceived exertion, and symptoms. The ECG recordings are stored electronically and can be shared over networks in large laboratories or clinics/hospitals. Disposable silver-silver chloride electrodes, prefilled with electrolyte paste or gel and having a strong adhesive backing to withstand the rigors of heavy exercise without losing skin contact, are generally employed for exercise electrocardiography. Recommendations for calibration and specifications for diagnostic ECG equipment have been published by the American Heart Association (Hellerstein, 1979).

In ECG terminology, a lead refers to two or more electrodes placed on the skin at specific anatomical locations and connected to the ECG computer by insulated wires. Single- and multiple-lead configurations are employed in exercise testing, and electrode placement varies accordingly. The anatomical locations of the electrodes for the various leads in the Mason-Likar (Mason and Likar, 1966) 12-lead system, the most common exercise lead system, are shown in Figure 6.1 and described in Table 6.1. It is very important that care be taken to place the electrodes at these anatomical landmarks, since failing to do

FIGURE 6.1 – (*a*) The chest electrodes are shown on a skeletal torso to illustrate the location of the electrodes relative to the intercostal spaces. (*b*) Electrode placement on a human torso for the Mason-Likar 12-lead configuration.

(*a*)
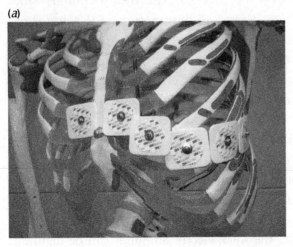

(*b*)
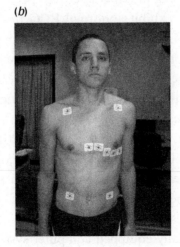

TABLE 6.1 – The anatomical location of the ECG electrodes for the Mason-Likar 12-lead configuration

ELECTRODE	ANATOMICAL LOCATION
Right Arm (RA)	The base of the right shoulder against the deltoid border about 2 cm below the clavicle but above border of pectoralis (in deltoid fossa).
Left Arm (LA)	The base of the left shoulder against the deltoid border about 2 cm below the clavicle but above border of pectoralis (in deltoid fossa).
Right Leg (RL)	Right anterior axillary line halfway between the costal margin and iliac crest.
Left Leg (LL)	Left anterior axillary line halfway between the costal margin and iliac crest.
V_1	Fourth intercostal space at right sternal border.
V_2	Fourth intercostal space at left sternal border.
V_3	Midway between positions for V_2 and V_4.
V_4	Fifth intercostal space at left midclavicular line.
V_5	Horizontal level of V_4 midway between V_4 and V_6 at left anterior axillary line.
V_6	Horizontal level of V_4 at left midaxillary line.

so will give ECG recording errors, and may result in diagnostic mistakes and inappropriate therapeutic interventions (Garcia-Niebla et al., 2009). The electrode connection arrangement and the polarity for the standard limb leads, the augmented limb leads, and the precordial (or chest) leads are shown in Table 6.2. When using automated equipment capable of recording the ECG from three leads simultaneously, it is standard practice to program the equipment to sequentially cycle every 2.5 seconds between groups of three leads, and to organize the paper printout into four columns on the recording paper. In column one, leads 1, 2, and 3 are shown; in column two, leads aVR, aVL, and aVF; in column three, leads V_{1-3}; and in column four, leads V_{4-6}. The ECG printouts that accompany this laboratory are labeled to show this standard format.

Studies have shown that the diagnostic usefulness of the exercise ECG increases when multiple lead configurations are employed, but that if only one lead can be used, V_5 will yield the most information related to heart blood flow during exercise (Gibbons et al., 1997; Robertson, Kostuk, and Ahuja, 1976; Shiran et al., 1997). A second popular single lead, especially if monitoring heart rhythm disturbances, is either a V_1 or a modified V_1 lead. In situations where more ECG information is desired during exercise testing, particularly in diagnostic settings, the Mason-Likar (Mason and Likar, 1966) modification of the conventional 12-lead

TABLE 6.2 – Electrode connection arrangement and polarity for Standard 12-lead ECG

CLASSIFICATION	LEAD	ELECTRODES CONNECTED	POSITIVE ELECTRODE
Standard Limb Leads	1	LA and RA	LA
	2	LL and RA	LL
	3	LL and LA	LL
Augmented Limb Leads	aVR	RA and (LA-LL)	RA
	aVL	LA and (RA-LL)	LA
	aVF	LL and (RA-LA)	LL
Precordial (Chest) Leads	V_1	V_1 and (LA-RA-LL)	V_1
	V_2	V_2 and (LA-RA-LL)	V_2
	V_3	V_3 and (LA-RA-LL)	V_3
	V_4	V_4 and (LA-RA-LL)	V_4
	V_5	V_5 and (LA-RA-LL)	V_5
	V_6	V_6 and (LA-RA-LL)	V_6

Note that the RL (right leg) electrode is the reference electrode for every lead configuration. LA = left arm, LL = left leg, RA = right arm, RL = right leg, V refers to precordial or chest leads.

configuration is routinely used. In this configuration, the limb electrodes are moved from the ankles and wrists to the base of the torso and shoulders, respectively (Figure 6.1). It should be noted that a rightward, inferior, posterior shift of the ECG axis may result from this modified electrode placement (Papouchado et al., 1987; Sevilla et al., 1989).

In Figure 6.2(*a*), an idealized ECG waveform is illustrated, and the commonly noted P, Q, R, S, and T wave components identified. Figure 6.2(*b*) illustrates the standard speed at which the ECG is recorded is 25 mm·sec^{-1}; therefore, 1 mm = 0.04 sec or 40 millisec. Using this information, heart rate and durations of various wave components can be determined in milliseconds. For example, one method of determining heart rate in beats per minute (bpm) from the ECG is to measure the R-wave to R-wave interval (R-R) in millimeters between two successive heart beats, then divide this value into 1,500. To illustrate this concept, assume that the measured R-R interval was 10 mm. The corresponding heart rate would be:

$$\text{Heart Rate}_{bpm} = 1500 \text{ mm} \cdot min^{-1} \div 10 \text{ mm} \cdot beat^{-1} = 150$$

Note from Figure 6.2(*b*) that a vertical deflection on the ECG paper corresponds to the voltage of the signal, either positive (upward) or negative (downward). It is standard procedure to calibrate the ECG recorder so that a 1 millivolt (mv) signal will cause a 10 mm (1 mm = 0.1 mv) deflection on the recording paper (Hellerstein, 1979). Therefore, the positive or negative voltage of any wave component can be determined by measuring its vertical deflection from the isoelectric line, also known as the ECG baseline. For example, it is very important in interpreting exercise ECGs to be able to determine the elevation or

FIGURE 6.2(a) – An idealized ECG waveform showing normal waves, durations, segments, and intervals.

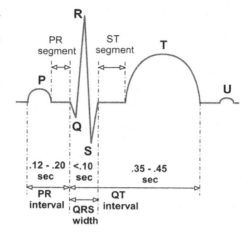

FIGURE 6.2(b) – A stylized illustration of an ECG waveform on ECG recording paper to show standard voltage and time orientations. The recording paper is marked with a grid in 1 mm squares.

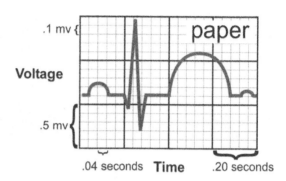

Paper speed = 25 mm / second
Heart Rate = number of R-waves in a 6 second strip divided by 10
= 1500 divided by the number of small boxes between consecutive R-waves
= large square estimation counts (300 - 150 - 100 - 75 - 60 - 50 - 43)

depression from baseline of the ST-segment. The normal orientation (up or down on the ECG recording paper), the voltage, and the duration of the waves, along with a summary of the cardiac events that normally occur in conjunction with each part of the waveform, are given in Table 6.3. It is important to understand that all of the specific waves that make up the entire ECG waveform, that is, the P, Q, R, S, and T waves, need not be found in every lead for the ECG to be interpreted as normal. For example, it is not abnormal for Q-waves to be absent from lead 1. On the other hand, the presence of large Q-waves in some leads, in leads 2, 3, and aVF may indicate the presence of a pathological condition, such as a myocardial infarction.

An understanding of the general pattern of heart tissue depolarization is required to interpret the 12-lead ECG. At rest, a small electrical potential difference exists across the membrane of heart cells, the inside charged negatively with respect to the outside. In this condition, the cells are said to be charged or polarized. When a cell is stimulated this electrical charge across the membrane is suddenly and temporarily reversed at the point of stimulation. This "depolarization" spreads wavelike over the cell membrane until the entire cell membrane is depolarized. The leading edge of this wave of depolarization carries a positive charge and the trailing edge carries a negative charge. To visualize this concept, think of a small traveling capsule containing a positive and negative charge, with the positive charge leading the way.

TABLE 6.3 – Durations and orientations of commonly identified waves and segments as found in the 12-lead system

WAVE/ INTERVAL	DURATION (SEC)	AMPLITUDE (MM)	USUAL WAVE ORIENTATION, IF PRESENT			CARDIAC EVENT
			Upright	Inverted	Variable	
P	<.11	<3	1, 2, aVF, V_{4-6}	aVR	3, aVL, V_{1-3}	Atrial Depolarization
P–R	.12–.20	isoelectric	–	–	–	Atrial activation A-V nodal delay A-V conduction
Q	<.03–.04	small <⅓ R	–	1-3, aVR-F, V_{5-6}	–	LV septal depolarization
R	–	variable	1-3, aVR-F, V_{4-6}	–	–	Mainly LV depolarization
S	–	variable	–	1-3, aVR-F, V_{1-2}	–	LV and RV depolarization
QRS	.05–.10	5–30	1, 2, aVF, V_{4-6}	aVR, V_{1-2}	3, aVL, V_3	Ventricular depolarization
S-T segment	–	–.5 to +1	–	–	–	Ventricular repolarization, slow phase
T	–	<5 limb <10 chest	1, 2, V_{4-6}	aVR	3, aVL, aVF, V_{1-3}	Ventricular repolarization, fast phase
Q-T	< ½ R-R	–	–	–	–	Duration ventricular systole

Caution: These criteria do not dogmatically hold in every circumstance. When questions arise, it is best to refer the ECG to an expert, such as a cardiologist.

This charged capsule is often referred to as a dipole. As shown in Figure 6.3, when this dipole is traveling through a heart muscle fiber *toward* a positive electrode, a positive (upward) deflection will be recorded. Conversely, when the dipole is moving *away* from a positive electrode, a negative (downward) deflection

FIGURE 6.3 – A representation of the QRS wave responses to an electrical dipole moving though heart muscle tissue. Panel A: A stimulus evokes depolarization at one end of the fiber. Both electrodes begin to write an upstroke as the positive (+) end of the dipole moves toward them. Panel B: As the dipole passes under and continues away from the left electrode leaving the negative end oriented toward the electrode, a downstroke is written followed by the beginning of a return to baseline. At the same time, the + end of the dipole is still approaching the right electrode, so the upstroke continues to climb higher. Panel C: The dipole has passed under the right electrode, and a downstroke ensues from this electrode followed by a return to baseline as the fiber is completely depolarized. When the fiber is repolarized the T-wave will be written.

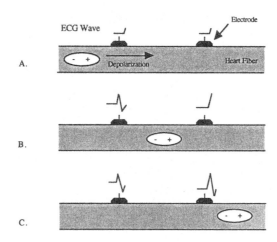

will be recorded. This concept is critical in understanding why the normal ECG waveforms look different in the various leads of the 12-lead system.

Of course, the heart is not a single muscle fiber but is composed of a multitude of fibers oriented in many different directions. As depolarization occurs concurrently in many fibers, the resulting dipoles will be moving simultaneously in many different directions in three-dimensional space (x-y-z axis). The summated effects of these multiple dipoles during any period of the cardiac cycle can be represented by a vector. A vector is a mathematical construct that indicates the magnitude and direction of a force; in this case, the summated or net electrical force in the heart muscle that occurs during depolarization. The vector corresponding to ventricular depolarization is called the mean QRS vector of the heart.

The orientation in two-dimensional space (x-y axis) of the mean QRS vector with respect to the frontal plane (front view) of the body describes what is known as the electrical axis of the heart. This electrical axis is found by determining the direction of the mean QRS vector, that is, the direction the "head" of the vector is pointing in the x-y axis of the frontal plane. Since the normal depolarization of the ventricles begins at the A-V node then travels through the bundle of HIS, bundle branches, Purkinje fibers, and ventricular muscle, the mean QRS vector usually points downward and to the left with respect to the frontal plane of the body (Figure 6.4).

Although such a general description of the location of the QRS mean vector is helpful, it is often more useful to quantify its orientation in space, since this more precise information can provide diagnostic clues related to heart problems such as heart attack and heart enlargement. As shown in Figure 6.4, the relative orientation of the limb leads creates a frontal plane reference system called a hexaxial array,

FIGURE 6.4 – The normal mean QRS vector and hexaxial array created by the combination of the 6 limb leads. The location of the positive (exploring) electrode for each lead is indicated by the oval labels containing the lead names.

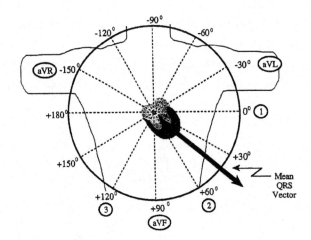

each line of which represents one of the six limb leads (1, 2, 3, aVR, aVL, aVF) oriented 30° apart. The electrical axis of the heart can be described and quantified using this array. Notice that lead 1 bisects the array configuration dividing it into upper and lower hemispheres, and lead aVF divides it into left and right hemispheres. Thus, the combination of these two leads divides the array into four quadrants. By convention, the positive pole of lead 1 is defined as 0° and the positive pole of aVF as +90°; leftward rotation from lead 1 (upward on the two-dimensional figure) is assigned negative values, and rightward rotation (downward) is assigned positive values (Dubin, 2000; Thaler, 2010).

The electrical axis can be placed in one of these quadrants quite easily by evaluating the net QRS deflection in each of these two leads. This is an application of the dipole principle described previously. Stated simply, the electrical vectors of the depolarizing heart muscle sum to an average vector, considered a wave of depolarization. If this wave is traveling toward a positive skin electrode, a positive (upward) deflection will predominate on the recorded ECG. Alternately, if the wave of depolarization is traveling away from a positive electrode, a negative (downward) deflection will predominate. This concept can be applied to evaluate the QRS voltage deflection of a representative ECG waveform in leads 1 and aVF, and the results used to locate the electrical axis of the heart in one of the four hexaxial array quadrants. This can be done by first adding the depth of the Q-wave to the depth of the S-wave below baseline in lead 1; this represents the negative voltage in lead 1. Subtract this sum from the height of the R-wave in lead 1, which is a measure of the positive voltage of the QRS waveform. The resultant value is the net QRS deflection, which can be positive, negative, or zero depending on the orientation of the mean QRS vector in the frontal plane of the body. Repeat this process for lead aVF. Use the net QRS deflection in both of these leads to place the heart axis in the hexaxial quadrant as follows (Dubin, 2000):

Positive in both leads	= **Normal Axis** (0° to +90°)
Lead 1 positive, aVF negative	= **Left Axis Deviation** (between 0° and −90°)
Lead 1 negative, aVF positive	= **Right Axis Deviation** (between +90° and +180°)
Negative in both leads	= **Indeterminate Axis** (between −90° and ±180°)

Although the methods will not be discussed in this laboratory exercise, the heart axis can be placed in the frontal plane with about ±15° accuracy. For now, it is sufficient to be able to place the axis in the proper quadrant.

It is not possible in this short introduction to describe all the details of ECG lead placements and interpretation. Many ECG interpretation concepts are summarized in the *Model for ECG Interpretation*, which accompanies this laboratory. For more complete and detailed information, many good ECG textbooks are available, some of which appear in the reference list for this laboratory (Dubin, 2000; Thaler, 2010).

Equipment Needed

1. 12-Lead ECG machine with paper recorder (calibrated (1 mV = 10 mm, paper speed 25 mm $\cdot sec^{-1}$)
2. Motor-driven treadmill or stationary cycle ergometer
3. Silver-silver chloride electrodes (disposable)
4. ECG prep abrading tape or gauze and alcohol
5. Electric hair trimmer or disposable dry razor

TECHNIQUES AND PROCEDURES

These procedures are those typically followed in research and clinical exercise testing laboratories, but may be applied anywhere exercise testing is combined with electrocardiography. They comply in general with the recommendations of the American College of Sports Medicine (Thompson, Gordon, and Pescatello, 2010), and the American Heart Association (Myers et al., 2009; Thompson, Gordon, and Pescatello, 2010). The focus is on the most important steps to ensure accurate recording of the ECG during rest and exercise. Since the Mason-Likar (Mason and Likar, 1966) configuration for the 12-lead ECG is almost universally employed when diagnostic information is required from a GXT, only this electrode preparation procedure will be described. Preparation for single lead configurations is the same, except that fewer electrodes are placed on the skin (generally 3), and the anatomical positions of the electrodes will vary from those described for the 12-lead ECG. **Before beginning, it is very important that the student review the ECG tracings at the end of this laboratory chapter to become familiar with the appearance and progressive changes of the ECG with exercise.**

PRELIMINARY PROCEDURES

These procedures can be completed by one technician, and this is usually the case in professional settings. But in this student laboratory, teams of three persons will work together, one person to serve as the subject, one to perform the duties of the ECG technician prepping the subject, and one to control the ECG/treadmill equipment and record data.

1. As noted in the introduction, the operational controls, commands, and features of ECG and treadmill equipment vary widely with make and model. The first step to begin this laboratory, and before a subject is readied for the procedure, is to become familiar with the operational controls and commands of the specific equipment available for use in the current facility.
2. Greet the subject at the ECG station in a calm, reassuring manner, and provide a simple explanation of the procedure. Ask the subject about any previous ECG experience and whether or not the ECG was normal. This information will help guide the extent of the procedure explanation provided to the subject, and will also help in interpreting the ECG in later steps.

3. Before beginning any laboratory procedures that involve potential risk to the subject, it is usually required that the subject (or patient) read and sign an informed consent. This should be done before proceeding if required (see the Informed Consent for Submaximal Exercise Testing in Laboratory 8 for an example).
   a. Because there is exercise involved in this laboratory experience, even though it is relatively of low intensity, in the process of obtaining informed consent, inquire about any exercise limitations known to the subject. These limitations will be considered later in a decision to allow or *not* allow the subject to exercise.
4. If the ECG system available for use is computerized, enter demographic and medical information about the subject now. It is important to enter this information accurately, as the subject's ECG file will be stored on the local ECG computer, in addition to the hard copies that will be recorded. In many hospitals and large laboratories, the ECG data are also stored on computer networks to be shared among medical care or research personnel. The data entry capabilities vary with the make and model of ECG equipment, but most allow the entry of the subject's name, age, date of birth, height, weight, gender, and ethnicity. Other important information often entered includes resting systolic and diastolic blood pressure, significant medical history, and any medications the subject may be taking. At the very least, every ECG recording printed should have the name of the subject and the date of the ECG test written on it. This should be done by hand if the ECG machine does not automate this function.

ELECTRODE PLACEMENT AND SKIN PREPARATION

Proper skin preparation is the essential first step for recording high-quality electrocardiograms during exercise. Failure to prepare the skin properly and consistently for an exercise test will result in an ECG signal that cannot be monitored or accurately interpreted because of artifact. The following procedure is recommended.

1. Ask the subject to undress to the waist and lie down on his or her back on the examination table, arms comfortably at the side, and legs uncrossed. For obvious reasons, it is preferred that women locate and place electrodes on women subjects whenever possible. In a nonmedical laboratory, women are usually asked to wear a two-piece bathing suit top or an exercise sports top under loose-fitting clothing. A lab gown may also be worn with a front opening for the ECG procedure. Underwire brassieres must not be worn.
2. Carefully review Figure 6.1(*a*), (*b*), and Table 6.1 to be sure you can accurately identify the general locations to be used for electrode placement. It is very important to carefully place the electrodes in the standard locations to ensure accurate interpretation of the ECG (Garcia-Niebla et al., 2009). A more detailed description of standard procedures for locating the ECG electrodes are published elsewhere (Rautaharju et al., 1998). In women and the obese where chest tissue may make precise placement of electrodes difficult, it is recommended that the electrodes be placed on the breast tissue rather than under the breast, and as closely as possible to the prescribed anatomical locations.
   a. Begin with carefully locating the right leg electrode. This is the electrical reference (ground) electrode for all 12 leads. If this electrode is improperly placed or prepped, it will negatively affect recordings from all leads. Mark the location with a felt-tip marker. Continue marking the locations of the right arm, left leg, and left arm electrodes. *Note:* As one becomes proficient at locating

FIGURE 6.5 – The subject's torso should resemble the torso in this picture after the locations of the electrodes have been marked.

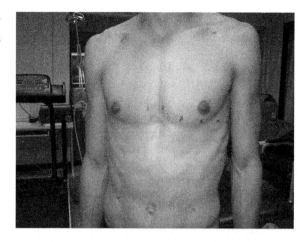

the ECG electrodes, marking the locations will not be necessary. See Figure 6.5 for a visual of a subject's torso marked for electrode application.

b. Next find and mark the location of the V_1 chest electrode.
   i. Stand on the left side of the subject and start by locating the suprasternal notch at the top of the manubrium, which is the upper portion of the sternum.
   ii. Firmly palpate the chest using the fingertips of the middle and index fingers to locate the subject's right first rib, which is just below the suprasternal notch and to the right of the sternum. Just down the chest from the first rib and adjacent to the right side of the sternum is the first intercostal space, the space between the first and second ribs.
   iii. Firmly press and hold the middle finger of your right hand in this first intercostal space, then with the index finger palpate down the chest to find the second rib. Drop your finger below the second rib and into the second intercostal space.
      1. An **alternate** procedure to locate the second rib is to firmly palpate the manubrium (sternum) to identify the sternal angle. This bony protrusion marks the point where the sternum and second rib meet. Firmly locate the second rib on the right sternal boarder, and drop your middle finger into the second intercostal space just below the second rib.
   iv. Continue down the rib cage palpating firmly with the fingertips by 'walking' the fingers down from the second intercostal space to locate the third and fourth ribs and intercostal spaces.
   v. Mark the fourth intercostal space at the right sternal boarder. This is the location for chest lead V_1.
c. Follow with locating and marking the V_2 electrode in the fourth intercostal space, left sternal border, at the same horizontal level as V_1.
d. Next locate and mark the V_4 electrode location.
   i. Palpate as before with the fingertips to find the fifth intercostal space on the left side of the sternum, just below the V_2 position.
   ii. Keep the fingertip in the fifth intercostal space and move it laterally and slightly diagonally downward to the subject's left. Stop at the point where the midclavicular line intersects the fifth intercostal space. The midclavicular line is an imaginary vertical line perpendicular to the midpoint of the clavicle measured from the manubrium to the left deltoid boarder. The

intersection of the midclavicular line and the fifth intercostal space is usually just below the left nipple. Mark this as the location for V_4.
   e. Now mark V_3 on the diagonal line midway between V_2 and V_4.
   f. Next mark V_6.
      i. Ask the subject to extend the left arm laterally away from the body. Identify the midaxillary line, which is halfway between the anterior and posterior axillary lines.
      ii. Follow the midaxillary line down the vertical plane of the thorax to the intersection of a horizontal line from the location of V_4. This intersection is the location for the V_6 electrode. Be careful not to place this electrode below a horizontal line from the V_4 electrode location.
   g. Finally mark the V_5 location on a horizontal line midway between V_4 and V_6, at the point of intersection of the anterior-axillary line.
3. Now that the ECG electrode locations have been marked, the next steps involve careful skin preparation of the sites and placement of good-quality electrodes. Failure to follow these steps will almost surely lead to uninterpretable ECG recordings during exercise.
   a. Electrode sites for exercise testing *must* be free from hair. Shave each site if necessary using an electric hair trimmer or disposable dry razor. Failure to remove hair from the electrode sites will lead to poor and uninterpretable ECG recordings.
   b. The skin is normally covered by a layer of oils and the superficial horny layer of the epidermis which will, if not removed, increase impedance (resistance) to electrical current flow across the skin-electrode interface resulting in poor-quality ECG recordings during exercise. To lower the skin impedance, the skin should be lightly abraded using gentle yet firm strokes with ECG abrasive prep tape or fine-grain sandpaper at the precise location where the center of the electrode will be placed. As an alternate procedure, rub each site firmly and briskly in a *vertical direction only* with an alcohol-soaked gauze swab. Many laboratories use both of these skin preparation techniques to ensure excellent ECG signals during exercise. The skin should turn slightly red if this procedure is performed properly.
   c. Apply the electrodes to the prepared sites. The electrolyte gel at the center of the electrode should be placed directly over the abraded area at each electrode site. Smooth around the adhesive part of the electrode with a finger to ensure good adhesion of the electrode to the skin, but do not to press on the center of the electrode. In some instances, especially when an individual is expected to sweat profusely, e.g., in athletic performance testing, it may be necessary to tape each electrode to the skin, or to wrap the chest and abdomen with an elastic wrap to secure the electrodes and minimize motion artifact.
   d. As a check on the quality of the ECG electrode preparation procedures, a test of the skin impedance at each electrode site should be carried out before exercise begins. Most modern ECG machines have the capability to test electrode impedance as a menu option, and this should routinely be done before exercise testing. Any electrodes that do not pass this test must be removed, the skin again prepped, and new electrodes placed before proceeding with the ECG recordings.
      i. In the absence of this option on the ECG equipment, a substitute test using a common volt-ohm meter can be employed to check the ohms of resistance of all electrodes against the right leg (reference or ground) electrode. If the resistance is greater than 5,000 ohms for any electrode, remove the offending electrode(s), repeat the skin prep procedure, place a new electrode at the site, and retest the resistance.

RECORDING THE RESTING SUPINE 12-LEAD ELECTROCARDIOGRAM

1. With the subject lying down, arms comfortably at the side, and legs uncrossed, attach the lead wires from the ECG recorder to the correct electrodes. CAUTION: Be certain to attach the correct lead wires from the ECG machine to the correct electrodes for each of the 12-leads (refer to Table 6.2). Failure to follow this simple directive will result in uninterpretable ECG recordings (Garcia-Niebla et al., 2009). Double-check all connections before proceeding.

2. Just prior to recording the resting ECG to paper, visually check each lead display on the monitor to be sure the signal is clear and free of artifact. Gently tapping on each electrode with a finger while viewing the monitor will usually produce excessive "noise" artifact in any lead in which the skin has not been properly prepped or the electrodes are defective. If there is excessive artifact detected, remove the offending electrode(s) and repeat the skin prep procedure. Most modern ECG machines also have a switch to electronically filter unwanted "noise" artifact out of the ECG signal. Engaging the filter usually helps provide a clearer signal and cleaner recording than would otherwise be possible.

3. Ask the subject to lie still without talking, and then toggle the switch on the ECG machine to record the 12-lead ECG to paper. Carefully review the printed tracing for artifact-free quality. Review the **RESTING SUPINE ECG** at the end of this laboratory chapter for an example of an acceptable recording. If the 12-lead ECG from the subject is deemed unacceptable, check all leads and electrodes, then record a second resting ECG. *Do not proceed beyond this point until a suitable resting 12-lead ECG is obtained.* A poor ECG signal at rest will almost always get worse with exercise, so it is very important that the resting ECG be free of artifact.

4. As noted above, most modern equipment can be programmed to automatically write essential information on every recorded ECG. Be sure that this resting, supine ECG is labeled as such, even if the information must hand-written on the 12-lead recording. At minimum, every ECG recorded to paper must have written on it the subject's name and date of testing, along with the subject's status when the ECG was recorded; e.g., *SUPINE REST* would describe this first 12-lead ECG. During the exercise test, the ECG recordings should be labeled with the exercise stage, speed, and grade. If blood pressures are taken, these should also be written on the ECG recordings.

INTERPRETATION OF RESTING ECG AND DECISION POINT

At this point in the procedure, a quick review of the resting 12-lead ECG should be performed to be sure there are no abnormal ECG findings and that it is safe for the subject to exercise. In a clinical setting, this evaluation of the resting ECG generally would be completed by the attending physician, or by a nurse or exercise physiologist trained in ECG interpretation. The following procedure is intended to aid the student in understanding the general procedure for ECG interpretation.

1. First, review the *Model for ECG Interpretation,* which is included at the end of this laboratory, along with the introductory material to this laboratory explaining some of the basic principles of electrocardiography. CAUTION: This interpretation model is *not* intended to give a complete system for ECG interpretation dogmatically followed in all circumstances. It is simply intended to provide the novice with an understanding of the standard sequence of ordered steps commonly included in any method of ECG interpretation.

2. Use the resting, supine 12-lead ECG recorded from your subject along with the *Model for ECG Interpretation* as a guide to complete the *Resting ECG Interpretation Worksheet* included at the end of this laboratory.
3. **DECISION POINT.** At this point in the procedure before beginning exercise, a decision must be made as to the presence of any factor or factors, commonly called contraindications, which would make exercise unsafe for the subject. As examples, factors like skeletal-muscle injuries or physical disabilities, metabolic disorders, or acute illnesses could put the subject at risk for injury if she or he were to exercise (Thompson, Gordon, and Pescatello, 2010). The most common ECG contraindications are shown in Table 6.4. If the ECG is thought to be abnormal, or if one of the *absolute* ECG contraindications is noted, the exercise test should not be performed and the subject should be referred to a physician for further evaluation before exercise.

TABLE 6.4 – ECG contraindications to exercise testing and conditions that may invalidate the diagnostic use of the exercise ECG

ECG CONTRAINDICATIONS TO EXERCISE TESTING
Absolute Contraindications
1. Uncontrolled heart arrhythmias accompanied by symptoms or compromised hemodynamics.
2. Recent change in resting ECG suggesting recent myocardial infarction (≤2 days), significant ischemia, or other acute cardiac events.
Relative Contraindications
1. High-degree atrioventricular block.
2. Tachyarrhythmias or bradyarrhythmias.
CONDITIONS WHICH MAY INVALIDATE THE DIAGNOSTIC USE OF THE EXERCISE ECG
1. Existence of resting repolarization abnormalities, such as left bundle branch block.
2. Pre-excitation syndromes, e.g., Wolf-Parkinson-White syndrome (WPW).
3. Left ventricular hypertrophy with ST strain pattern.
4. Medications (e.g., Digoxin) which alter the ST-T wave.
5. Musculoskeletal or other factors limiting exercise capacity to below ischemic threshold.
6. Electrolyte disorders, e.g., hypokalemia.
7. Mitral valve prolapse.

Adapted from Thompson, Gordon, and Pescatello, 2010, and Gibbons et al, 1997.

EXERCISE ECG

1. The procedures that follow are written primarily for a treadmill exercise test, since the treadmill is the predominate ergometer in use for exercise testing. If a cycle ergometer is used, the pre-exercise ECG should be recorded with the subject seated on the cycle ergometer, but the timing of ECG recordings and stage times will remain the same as for the treadmill.
2. Before beginning, review again the Stage 1, 2, and Recovery ECGs at the end of this laboratory chapter as a reminder of the acceptable appearance of exercise electrocardiograms. Keep these at hand for reference during the exercise portion of this laboratory.
3. **IMPORTANT PROCEDURAL ISSUES:**
    a. **During exercise, it is critical that the technician(s) conducting the exercise test closely observes the ECG monitor for:**
        i. **Any abnormal ECG responses which might indicate cardiac problems. If one of the ECG Criteria for Stopping the Graded Exercise Test (Table 6.5) is noted, the exercise should be stopped immediately!**
        ii. **Excessive artifact. Correct loose lead wires or electrodes as needed to improve the ECG signal. Often this can be done while the subject continues to exercise. If not, and the ECG continues to be uninterpretable owing to excessive artifact, the exercise should be stopped and the problem corrected before proceeding.**
    b. **Maintain visual assessment of the subject and communicate with the subject throughout the exercise test.**
    c. **Stop the test immediately if the subject requests to do so.**

TABLE 6.5 – ECG criteria for stopping the graded exercise test

Absolute Criteria
1. Sustained ventricular tachycardia.
2. +1-mm ST-segment elevation in leads without significant Q-waves, other than leads V_1 or aVR.
3. Technical difficulties monitoring the ECG.
Relative Criteria
1. Exercise-induced bundle branch block indistinguishable from ventricular tachycardia.
2. ST-depression exceeding 2-mm horizontal or downsloping.
3. Any arrhythmia other than sustained ventricular tachycardia, such as heart block, bradyarrhythmias, multiformed or R-on-T premature ventricular contractions (PVC), or increasing frequency of PVCs.
4. Marked QRS axis shift.

Adapted from Gibbons et al., 1997, and Thompson, Gordon, and Pescatello, 2010.

4. Most computerized ECG and treadmill (or electronically braked cycle ergometer) systems can be programmed to automatically control the stage time, workload, and ECG printouts. If this is possible on the ECG machine available to use, check to be sure the treadmill protocol is set to "Bruce," or the appropriate custom protocol for this laboratory exercise. If a cycle ergometer protocol will be used, the system may have to be programmed for the protocol prescribed in Table 6.6, or the equipment may be operated manually. Programmed correctly, the ECG machine will advance the exercise stages as prescribed in Table 6.6, which are the first two stages of the standard Bruce protocol (Bruce, Kusumi, and Hosmer, 1973), and will automatically print an ECG during the last 10 seconds of each exercise stage, at the end of the exercise protocol, and the last few seconds of recovery.
5. Ask the subject to stand, attach the ECG lead harness securely to the waist, move to the treadmill (or cycle ergometer), and stand (sit, if cycle ergometer) quietly with arms relaxed at the side.
   a. Toggle the 12-lead switch or computer key to record to paper the standing rest, pre-exercise 12-lead ECG. If a cycle ergometer is used for the GXT, record the subject's pre-exercise ECG with him or her seated quietly on the ergometer.
   b. Review the recorded 12-lead, and if excess artifact is present repeat the recording. Do not proceed to exercise until a high-quality pre-exercise ECG is recorded.
   c. Label this ECG as *STANDING REST* (or *SEATED* if cycle ergometer). Label this and all subsequent ECG recordings, by hand if not automated by the ECG machine, with the subject's name, date, exercise stage, exercise time, and workload (i.e., speed and grade of the treadmill or rpm and watts of the cycle ergometer).
6. Thoroughly explain the testing procedure to the subject, including the exercise stage changes in workload to be expected each 3 minutes. Ask the subject to step off the treadmill, and then start the treadmill running slowly, 0% grade, 1.5 mph. Demonstrate to the subject how to safely step onto the moving treadmill.
   a. First, step up and straddle the slowly moving belt while using the hands to grasp the safety handrails of the treadmill.
   b. From this straddle position, lift one leg and gently "paw" the belt to get the feel of the belt speed.
   c. Keep holding the safety rails, step on the moving belt, and begin walking. Once comfortable with the speed, release the hands from the safety rails and begin walking normally on the moving belt.
   d. Step off the treadmill and turn off the treadmill belt.

TABLE 6.6 – Submaximal exercise protocol to be followed in this laboratory

TREADMILL				CYCLE ERGOMETER			
Stage	Time (min)	Velocity (mph)	Grade %	Stage	Time (min)	RPM	Work (Watts)
1	3	1.7	10	1	3	50	50
2	3	2.5	12	2	3	50	100
Recovery	2	1.7	0	Recovery	0	50	25

7. Ask the subject to step onto the treadmill, grasp the safety handrails, and straddle the belt.
    a. After carefully checking to be sure the subject is completely off the treadmill belt, start the belt at 0% grade and 1.5 mph. Coach the subject through the safe method just demonstrated to step on the moving treadmill and walk.
    b. It is recommended when testing older and more unstable subjects the technician place a hand on the back of the subject to protect against an accidental fall.
    c. As soon as the subject is comfortably walking, ask them to release the support rails and walk naturally.
    d. Verbally alert the subject that the exercise test is about to begin, and then toggle the switch or computer key to automatically (or manually) initiate Stage 1 of the exercise protocol as shown in Table 6.6.
8. Record a 12-lead ECG during the last 10 seconds of Stage 1.
    a. Visually scan the ECG as it is printed for excessive artifact. Compare to the Stage 1 example at the end of this laboratory chapter. If excessive artifact is noted, correct the cause and record a second ECG as soon as possible. Label this second ECG appropriately, especially noting the time of the recording relative to the exercise protocol.
    b. Be sure the ECG is labeled with the subject's name, date, stage number, speed, grade, and heart rate.
9. After three minutes in Stage 1, alert the subject and then advance the workload to Stage 2 for an additional 3 minutes. Do this manually if the task is not performed automatically by the equipment.
10. Record a 12-lead ECG during the last 10 seconds of Stage 2 and label appropriately. Compare with the Stage 2 ECG included at the end of this chapter. Correct any cause of excessive artifact, and record another ECG as soon as possible if necessary.
11. At the end of 3 minutes in Stage 2, alert the subject and then return the treadmill grade to 0% and the speed to 1.7 mph for the specified 2-minute walking recovery. In the case of a cycle ergometer, reduce the workload as shown in Table 6.6. **DO NOT STOP THE TREADMILL OR CYCLE ERGOMETER** at this time, since an active exercise recovery is prescribed in this and most exercise test protocols.
    a. On automated systems, a keystroke command to *End Exercise* (or a command similarly titled) will accomplish this, and will also record a 12-lead ECG at this time. This 12-lead recorded at the end of exercise should be labeled *MAX EXERCISE* or *END EXERCISE*. (Note, since Stage 2 and the exercise test ended simultaneously in this submaximal protocol, the Stage 2 and MAX EXERCISE ECG will be very similar. In practice, it is seldom that an exercise stage and maximal effort end simultaneously.) If the system used is not automated, the operator will be required to manually perform these end exercise tasks.
12. Instruct the subject to continue exercising through a 2-minute recovery stage. Record a final ECG during the last 10 seconds of recovery and label this ECG as with the recovery time, speed, and grade. Compare with the Recovery ECG included at the end of this chapter.
13. At the end of recovery, inform the subject the test is completed and then toggle the equipment or computer key to stop the treadmill.
14. Disconnect the ECG lead wires and dock them on the ECG machine. The subject can now be dismissed.

15. Use the exercise ECG printouts recorded from the subject to complete the *Exercise and Recovery ECG Interpretation Worksheet* found at the end of this laboratory chapter. As with the resting ECG, refer to the *Model for ECG Interpretation* when interpreting these ECGs. See Table 6.7 for the most common normal and abnormal ECG responses to exercise.

TABLE 6.7 – Possible ECG responses to exercise

NORMAL ECG RESPONSES TO EXERCISE
1. Shortened Q-T interval.
2. Reduced R-wave amplitude, especially over the left ventricle.
3. J-point may be displaced below isoelectric line.
4. May be a positive upslope on the ST-segment returning to isoelectric by .08 sec past J-point.
5. Linear increase in heart rate with increased workload; in untrained about 10 bpm increase per MET.
6. Possible isolated atrial ectopic beats and unsustained superventricular tachycardia.
ABNORMAL ECG RESPONSES TO EXERCISE
1. Horizontal or downsloping ST-segment depression exceeding 1 mm (.1 mv) extending for .08 sec past the J-point.
2. ST-segment elevation exceeding 1 mm.
3. Increased R-wave amplitude.
4. Sustained supraventricular tachycardia.
5. Ventricular tachycardia defined as at least 3 successive PVCs.
6. Increase in ventricular arrhythmias (PVCs), especially if multiformed PVCs during exercise or recovery.
7. Blunted heart rate recovery, ≤12 bpm
8. Chronotropic incompetence on maximal GXT defined as maximal exercise heart rate at volitional exhaustion more than 2 standard deviations below age-predicted heart rate (subject not taking ß-blocking medications)

Adapted from Thompson, Gordon, and Pescatello, 2010.

QUESTIONS AND LEARNING ACTIVITIES

1. On the SUPINE REST ECG, label the P, Q, R, S, and T waves, the PR interval, and the ST segment in leads 2, aVR, V_1, and V_5.
   a. Measure the voltage and duration values for each of these waveform components in lead 2. Provide a comparison to the normal ranges for these ECG wave component parts.
   b. Describe the cardiac events, which correspond to each of these ECG wave components.
2. Illustrate and explain the reasons for a change in the waveform recorded from leads 1 and 3 if the right arm and left arm lead wires were mistakenly reversed during the ECG prep procedure.
3. Describe the main orientation (up or down) of the QRS complex in each of the leads 1, 3, and aVR if the direction of the mean electrical axis of the heart pointed upward and toward the left arm. (*Hint:* Consider the hexaxial array for orientation of the frontal plane leads.)
4. Compare the 12-lead ECG labeled *MAX EXERCISE (or END EXERCISE*) recorded from the subject with the EXERCISE ECG included at the end of this chapter from a 55 year-old man who had high blood pressure. Provide a description of the ECG features that differ, if any, between these two exercise electrocardiograms, particularly focused on the heart rate, ST segment, and T-waves. Are both normal? Provide reasons why or why not. (Evaluate using the *Model for ECG Interpretation at the end of this laboratory chapter.*)

QUESTIONS AND LEARNING ACTIVITIES FOR GOING FURTHER

1. What changes in the resting ECG would you expect to see if a subject told you they had been diagnosed with sinus tachycardia and a right bundle branch block?
2. How would you recognize ventricular tachycardia on the exercise ECG? Why is this a reason to stop the exercise test?
3. Illustrate and explain the changes you might note on a resting ECG from 22-year-old male marathon runner. Why would these changes not be considered symptoms of disease?

REFERENCES

Bruce, R. A., F. Kusumi, and D. Hosmer. "Maximal Oxygen Intake and Nomographic Assessment of Functional Aerobic Impairment in Cardiovascular Disease." *Am. Heart J.* 85 (1973): 4, 546–62.

Crouse, S. F., T. Meade, B. E. Hansen, J. S. Green, and S. E. Martin. "Electrocardiograms of Collegiate Football Athletes." *Clin. Cardiol.* 32 (2009): 1, 37–42.

Dubin, D. *Rapid Interpretation of EKG's : An Interactive Course.* Tampa, FL.: Cover Publishing, 2000.

Garcia-Niebla, J., P. Llontop-Garcia, J. Ignacio Valle-Racero, G. Serra-Autonell, V. N. Batchvarov, and A. Bayes de Luna. "Technical Mistakes during the Acquisition of the Electrocardiogram." 14 (2009): 4, 389–403.

Gibbons, R. J., G. J. Balady, J. W. Beasley, J. T, Bricker, W. F. C. Duvernoy, V. F. Froelicher, D. B. Mark, T. H. Marwick, B. D. McCallister, P. D. Thompson, W. L. Winters, and F. G. Yanowitz. ACC/AHA Guidelines for Exercise Testing—A report of the American College of Cardiology American Heart Association task force on practice guidelines (committee on exercise testing). *J. Am. Coll. Cardiol.* 30 (1997): 1, 260–311.

Hellerstein, H. K. "Specifications for Exercise Testing Equipment—American Heart Association Subcommittee on Rehabilitation Target Activity Group." *Circulation* 59 (1979): 4, A849–A854.

Mason, R. E., and I. Likar. "A New System of Multiple-Lead Exercise Electrocardiography." *Am. Heart J.* 71 (1966): 2, 196–205.

Myers, J., R. Arena, B. Franklin, I. Pina, W. E. Kraus, K. McInnis, G. J. Balady. "Council Clin Cardiology and Council Cardiovasc Nursing: Recommendations for Clinical Exercise Laboratories—A Scientific Statement From the American Heart Association." *Circulation* 119 (2009): 3144–61.

Papouchado, M.. P. R. Walker, M. A. James, and L. M. Clarke. "Fundamental Differences between the Standard 12-Lead Electrocardiograph and the Modified (Mason-Likar) Exercise Lead System." *Eur. Heart J.* 8 (1987): 7, 725–33.

Rautaharju, P. M., L. Park, F. S. Rautaharju, and R. Crow. "A Standardized Procedure for Locating and Documenting ECG Chest Electrode Positions: Consideration of the Effect of Breast Tissue on ECG Amplitudes in Women." *J. Electrocardiol.* 31 (1998): 1, 17–29.

Robertson, D., W. J. Kostuk, and S. P. Ahuja. "Localization of Coronary-Artery Stenoses by 12 Lead ECG Response to Graded Exercise Test: Support for Intercoronary Steal." *Am. Heart J.* 91 (1976): 4, 437–44.

Sevilla, D. C., M. L. Dohrmann, C. A. Somelofski, R. P. Wawrzynski, N. B. Wagner, and G. S. Wagner. "Invalidation of the Resting Electrocardiogram Obtained via Exercise Electrode Sites as a Standard 12-Lead Recording." *Am. J. Cardiol.* 63 (1989): 1, 35–39.

Shiran, A., D. A. Halon, A. Laor, A. Merdler, E. Karban, and B. S. Lewis. "Exercise Testing in Patients with Chest Pain and Normal Coronary Arteries: Improving Test Specificity by Use of a Simple Logistic Model." *Cardiology* 88 (1997): 5, 453–59.

Thaler, M. S. *The Only EKG Book You'll Ever Need,* 6th ed. Philadelphia: Lippincott Williams & Wilkins, 2010.

Thompson, W. R., N. F. Gordon, and L. S. Pescatello (eds). *ACSM's Guidelines for Exercise Testing and Prescription.* Philadelphia: Lippincott Williams & Wilkins, 2010.

Model for ECG Interpretation

Interpretation of an ECG is best accomplished by following an ordered sequence of steps so that no important information is overlooked. The following model is not exhaustive but includes an analysis the most pertinent information and some of the most probable conclusions to be drawn when interpreting an ECG.

I. **Heart Rate**
   A. Heart rate can be determined using one of the following methods:
      1. Count number of complete ECG complexes in 6 seconds, including estimated fractions of cycles, then multiply by 10 to convert to beats per minute; e.g., 6.5 complexes × 10 = 65 bpm.
      2. Count the number of millimeters separating successive R-waves (or P-waves) and divide into 1,500.
   B. Decisions/conclusions
      1. Above 100 bpm at rest? ⇒⇒ **Tachycardia**
      2. Below 60 bpm at rest? ⇒⇒ **Bradycardia**
      3. 60–100 bpm? ⇒⇒ **Normal resting rate**
      4. 100–210 bpm with exercise? ⇒⇒ **Usually normal response**

II. **Heart Rhythm**
   A. Visually scan complete tracing for pauses, irregular rhythm, abnormal waveforms.
   B. Decisions/conclusions
      1. QRS following every P-wave?
         a. Yes/normal rate? ⇒⇒ **Sinus rhythm**
         b. No? ⇒⇒ **Nonconducted PAC or 2nd or 3rd degree A-V block**
      2. P-wave preceding every QRS?
         a. Yes/normal rate? ⇒⇒ **Sinus rhythm**
         b. Yes/tachycardia? ⇒⇒ **Normal with exercise or Supraventricular tachycardia**
         c. No, or abnormal P-wave? ⇒⇒ **PVCs, A-V nodal rhythm, A-V blocks**
      3. Early beats?
         a. Yes?
            (1) P-wave present? ⇒⇒ **PAC**
            (2) P-wave absent? ⇒⇒ **PVC**
      4. Pauses?
         a. Yes? ⇒⇒ **Sinus arrhythmia, nonconducted PAC, or 1st or 2nd degree A-V block**
      5. Total irregularity?
         a. Yes? ⇒⇒ **Atrial or ventricular fibrillation**

III. **Evaluate Conduction by Examining Intervals and Durations**
   A. P-R Interval
      1. Measure from beginning of P-wave to beginning of R-wave.
      2. Decision/conclusion
         a. Normal? ⇒⇒ **Normal A-V conduction**
         b. Short? ⇒⇒ **Preexcitation syndromes (WPW or LGL)**
         c. Long? ⇒⇒ **First degree A-V block**

B. QRS Interval
1. Measure width of QRS complex.
2. Decision/conclusion
   a. Normal? ➡➡ **Normal ventricular activation**
   b. Long & normal rate? ➡➡ **Bundle branch block**
   c. Long & tachycardia? ➡➡ **Ventricular tachycardia or Supraventricular tachycardia with aberrancy**

C. Q-T Duration
1. Measure from the beginning of the QRS complex to the end of the T-wave.
2. Decision/conclusion
   a. At rest, <½ preceding R-R? ➡➡ **Normal**
   b. Short? ➡➡ **Digitalis, calcium, or potassium excess**
   c. Long? ➡➡ **Decreased calcium or potassium**

IV. QRS Axis in Frontal Plane
A. For general axis determination, subtract the depth of the Q- and S-wave from the height of the R-wave in leads 1 and aVF to obtain the net QRS deflection, either positive or negative, in each of these two leads.
B. Decision/conclusion
   1. QRS positive in both leads? ➡➡ **Normal axis**
   2. Lead 1 positive, aVF negative? ➡➡ **Left axis deviation**
   3. Lead 1 negative, aVF positive? ➡➡ **Right axis deviation**
   4. QRS negative in both leads? ➡➡ **Indeterminate axis**

V. Hypertrophy
A. **Atrial** right or left (RAH or LAH).
   1. Check P-wave shape in lead 2 and V_1, measure amplitude and duration.
   2. Decision/conclusion
      a. Increased amplitude, peaked? ➡➡ **RAH**
      b. Increased width, biphasic V_1? ➡➡ **LAH or disease**
B. **Ventricular** right or left (RVH or LVH).
   1. Evaluate QRS in leads V_1 and V_{5-6}.
   2. Decision/conclusion
      a. R-wave >5 mm tall in V_1? ➡➡ **RVH**
      b. Sum of S-wave in V_1 and R-wave in V_{5-6} ≥ 35 mm. ➡➡ **LVH**

VI. **Ischemia/Infarction**
A. Evaluate Q-waves, ST-segments, and T-waves in all leads.
B. Decision/conclusion
   1. Abnormal Q-waves? ➡➡ **Old MI**
   2. Elevated ST-segment? ➡➡ **Acute MI, severe ischemia, or early repolarization**
   3. Depressed ST-segment? ➡➡ **Ischemia**
   4. Inverted T-wave? ➡➡ **May be ischemia or nondiagnostic**

FIGURE 6.6(A) – Resting supine ECG.

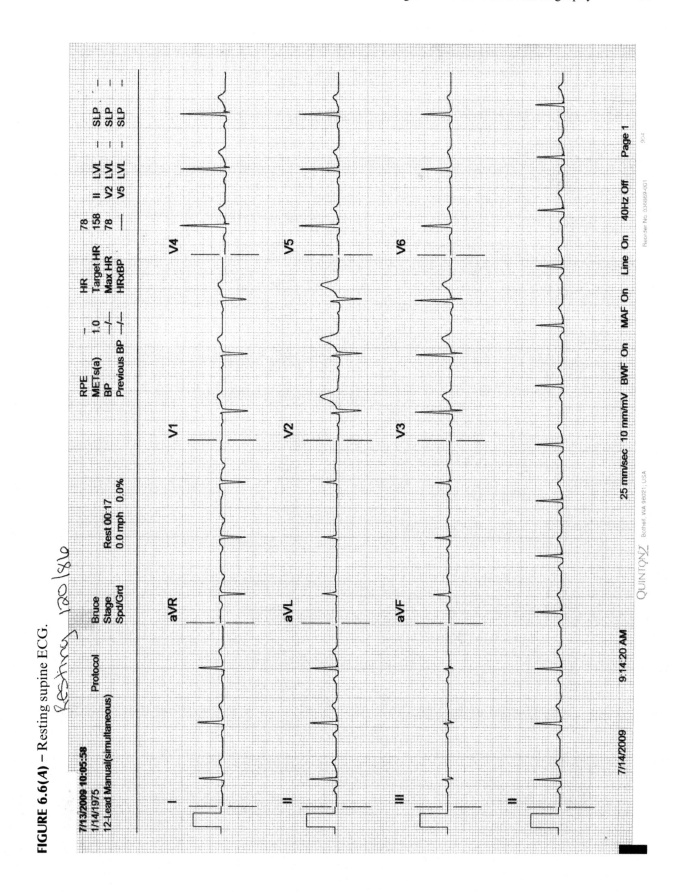

FIGURE 6.6(B) – Standing rest ECG.

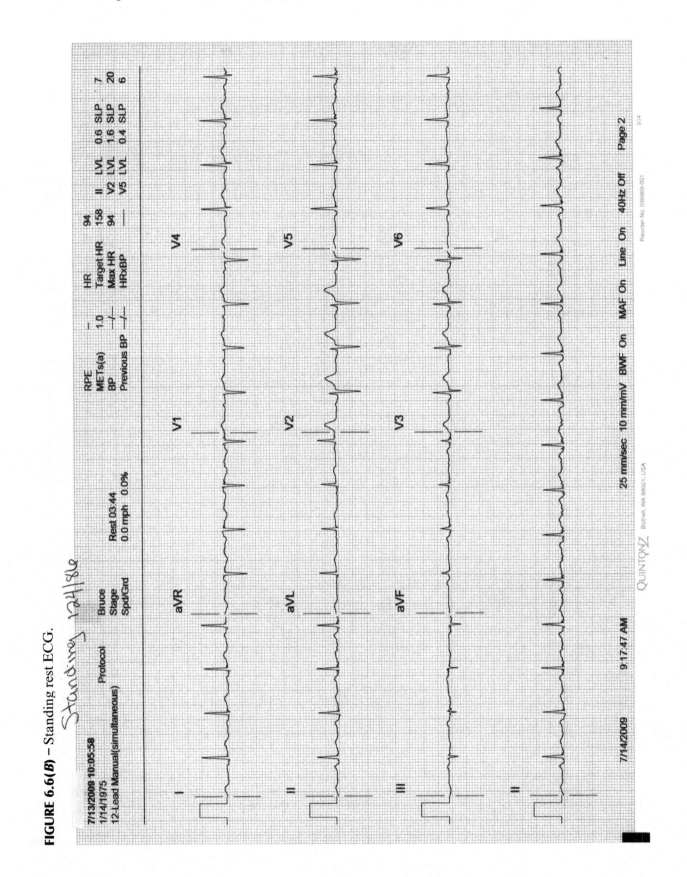

FIGURE 6.6(C) – Exercise Test Stage 1 ECG.

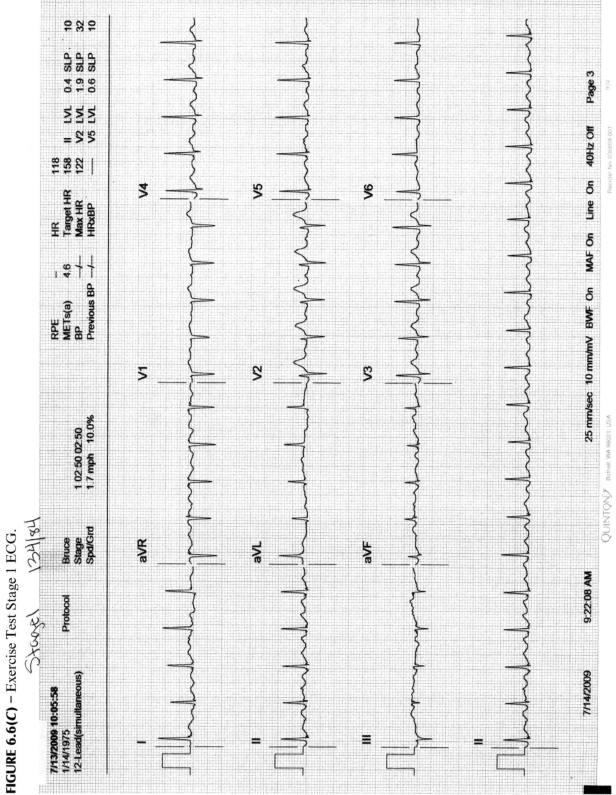

102 Resting and Exercise Electrocardiography

FIGURE 6.6(D) – Exercise Test Stage 2 ECG.

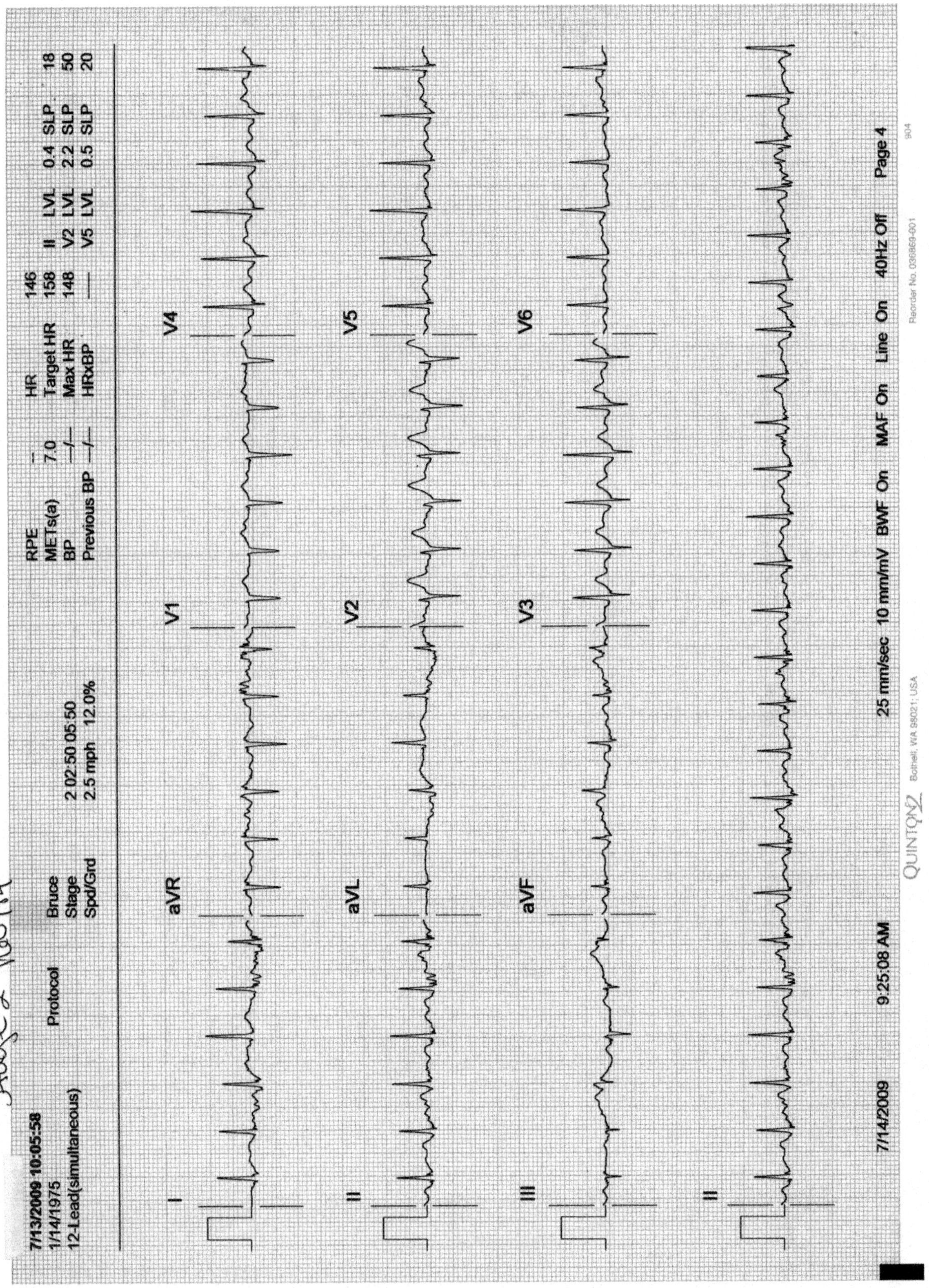

FIGURE 6.6(E) – Three-minute recovery ECG.

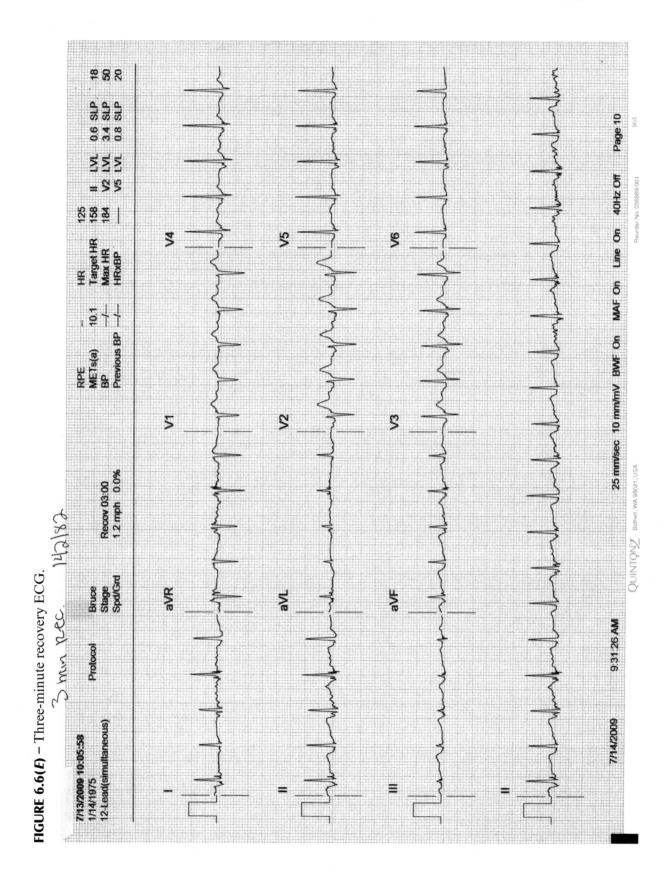

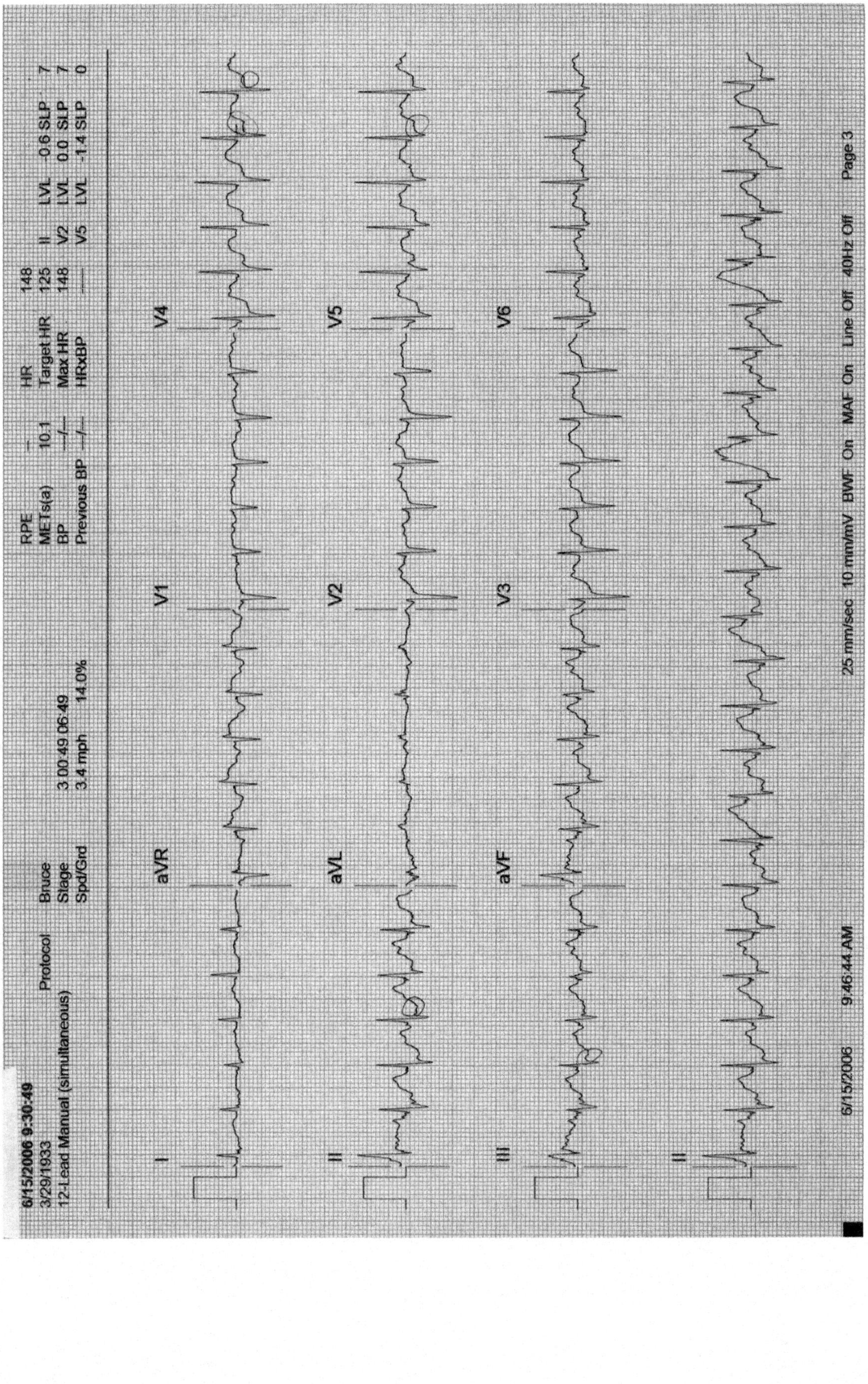

FIGURE 6.7 – Exercise ECG from a 55-year-old man with history of high blood pressure.

RESTING ECG INTERPRETATION WORKSHEET

Subject name: _____ Age: _____ Gender: _____ Date: _____

Heart Rate and Rhythm

Atrial rate: _____ bpm   Ventricular rate: _____ bpm

Rhythm description: _____

Conclusion: _____

Conduction Intervals

P-R: _____ sec   QRS: _____ sec   QT: _____ sec

Conclusion: _____

QRS Axis

Normal (0 to +90°): _____   LAD (0 to −90°): _____

RAD (+90 to +180°): _____   Indeterminate (−90 to ±180°): _____

Hypertrophy

<u>P-wave:</u>   Amplitude: _____ mm   Duration: _____ sec

<u>QRS:</u>   R-V_1: _____ mm   S-V_1 + R-V_5: _____ mm

Conclusion: _____

Ischemia/Infarction (Identify leads in which found, and measure of ST ↑ or ↓)

Significant Q-waves: _____   Abnormal T-waves: _____

Significant ST ↑ or ↓: _____

Conclusion: _____

General Conclusions: _____

EXERCISE AND RECOVERY ECG INTERPRETATION WORKSHEET

Exercise—Stage 1

Heart Rate: _____ **Rhythm:** _____

Change in Rhythm from Rest: _____

QRS interval: _____ sec **Change in R-Wave Amplitude from Rest:** _____

ST- T-waves: T-wave inversion: _____ Leads: _____

ST ↑ or ↓: _____ mm Shape: _____ Leads: _____

Conclusions/Comments: _____

Exercise—Stage 2

Heart Rate: _____ **Rhythm:** _____

Change in Rhythm from Rest: _____

QRS interval: _____ sec **Change in R-Wave Amplitude from Rest:** _____

ST- T-waves: T-wave inversion: _____ Leads: _____

ST ↑ or ↓: _____ mm Shape: _____ Leads: _____

Conclusions/Comments: _____

Recovery and Conclusion from Exercise

Heart Rate: _____ **Rhythm:** _____ **QRS interval:** _____ sec
ST- T-waves: T-wave inversion: _____ Leads: _____

ST ↑ or ↓: _____ mm Shape: _____ Leads: _____

Conclusions/Comments: _____

CHAPTER 7

MEASURING RESTING AND EXERCISE BLOOD PRESSURE

OUTCOME OBJECTIVES

Following successful completion of this exercise, the student should be able to:

1. Characterize the Korotkoff sounds associated with systolic and diastolic blood pressure.
2. Describe the proper procedures for measuring blood pressure.
3. Accurately measure resting and exercise blood pressure.
4. Explain the normal effects of exercise on systolic and diastolic blood pressure.

INTRODUCTION

The rhythmic contraction of the ventricular muscle of the heart exerts a force on the blood contained in the ventricular chambers, thereby producing the driving pressure for the vascular system. When the pressure inside the chambers exceeds that in the arteries, blood is forced through the heart valves (pulmonic and aortic) into the pulmonary and systemic vessels and flows from regions of relatively higher to lower pressure. The flow of blood is circuitous, passing down a pressure gradient through the arteries, arterioles, capillaries, venules, veins, and finally back to the right and left atria of the heart to begin the cycle again. Clearly, in the absence of a sufficient driving pressure, blood would not circulate. This, of course, is a problem of paramount importance when the death of heart muscle due to a myocardial infarction renders the pumping action of the heart ineffective, and blood circulation becomes inadequate to support life.

Blood pressure may be defined as the pressure exerted by the blood against the walls of the blood vessels. Since the heart is a pulsatile pump contracting during systole and relaxing during diastole an average of 70 times per minute, blood pressure in the arteries is constantly changing throughout the duration of each cardiac cycle, oscillating between a relatively higher and lower pressure. The higher pressure measurement is termed systolic blood pressure (SBP), which results from the systolic (contraction) phase of the cardiac cycle when the heart ejects a volume of blood into the aorta, called stroke volume. When the heart relaxes during the diastolic phase of the cardiac cycle, diastolic blood pressure (DBP) can be measured. Resting systolic and diastolic blood pressures are measured in units of millimeters of mercury (mmHg), and normally range from about 110–140 and 60–80 mmHg, respectively.

Pulse pressure (PP) and mean blood pressure (MBP) are two other terms frequently encountered in discussions of blood pressure, which are important to understand. Pulse pressure is defined as the difference between SBP and DBP. For example, if the SBP were 120 mmHg and the DBP were 80 mmHg, the PP would be equal to 40 mmHg. Mean blood pressure is the average blood pressure that occurs in the arteries over the duration of the cardiac cycle, and is an important measure of the true driving pressure for the circulatory system. It is difficult to measure without using an intra-arterial catheter and pressure transducer, but can be estimated from DBP and PP as follows:

$$\text{MBP}_{(mmHg)} = \text{DBP}_{(mmHg)} + (\text{PP}_{(mmHg)} \div 3)$$

The importance of blood pressure, and especially MBP, is best understood in light of its relationship to blood flow and resistance. Blood flow through any vascular bed is affected by two primary factors. First, a force of resistance is exerted by friction between moving blood cells and the blood vessel walls, which impedes blood flow through the vessels. This resistance is a function of the radius and length of the vessels through which the blood must pass, and the viscosity (thickness) of the blood. Since blood viscosity and the vessel length are not easily altered in the short term, changes in resistance that occur quickly are brought about by rapid changes in the radius of certain muscular blood vessels called resistance vessels, which are under regulatory control by a number of constriction and dilatory factors. The rate of blood flow through any vascular bed is inversely related to the magnitude of the resistance; that is, if resistance is increased, blood flow will be decreased. Since fluid dynamic studies in blood vessels show that resistance is proportional to the vessel radius raised to the fourth power (r^4), a relatively small change in the radius of a vessel results in a proportionally large change in resistance to blood flow.

Secondly, the rate of blood flow is directly proportional to the magnitude of the pressure gradient that exists across any vascular bed. In the whole body, this pressure gradient is related to the MBP in the arteries. This means that an increase in blood pressure, specifically MBP, will result in an increase in blood flow, as long as resistance does not change. If resistance increases in any vascular bed, then MBP must be raised to maintain the same level of blood flow, or flow will be proportionally reduced. The relationship between blood pressure, flow, and resistance can be expressed mathematically as:

$$\text{Blood flow}_{(L \cdot min^{-1})} = \text{MBP}_{(mmHg)} \div \text{Resistance}_{(mmHg \cdot min \cdot L^{-1})}$$

Rearranging this equation illustrates another aspect of this relationship which will be important in understanding the changes in blood pressure that occur with exercise.

$$\text{MBP}_{(mmHg)} = \text{Blood flow}_{(L \cdot min^{-1})} \times \text{Resistance}_{(mmHg \cdot min \cdot L^{-1})}$$

This shows that MBP will increase as the blood flow through the vessels increases. In the whole body, total blood flow per minute is equivalent to the cardiac output of the heart. Resistance is equivalent to the total peripheral resistance (TPR), also called systemic vascular resistance, which is defined as the sum of the resistance to blood flow in all the vessels throughout the body. Thus, when considering the whole body, cardiac output and TPR can be substituted for blood flow and resistance, respectively, in the equations above. During rest, cardiac output equals about 5 liters of blood per minute ($L \cdot min^{-1}$). During exercise, the demand for blood flow to the exercising muscles is greatly increased. In response, the heart contracts more rapidly and forcefully and ejects a larger stroke volume with each beat, which results in an increased cardiac output proportional to the intensity of the work. During very intense exercise, cardiac output may rise to a maximum value of 20 $L \cdot min^{-1}$ or more, a fourfold increase over resting values. As a consequence of the greatly increased flow, blood pressure in the arteries rises. Therefore, systolic blood pressure is affected most by stroke volume and cardiac output, increasing in direct proportion to exercise intensity, often reaching values of 180–200 mmHg at maximum exercise. Indeed, a fall in SBP of 10 mmHg or more during exercise (after the initial warm-up) is an ominous signal of a failing heart pump. On the other hand, it is unusual for SBP to rise above 250 mmHg during exercise, and exercise should be stopped when such pressures are measured. Such an excessive rise is sometimes found in individuals who have hypertension, or, it may be a predictor of the development of future hypertension in those with normal resting blood pressure (Gibbons et al., 1997; Thompson, Gordon, and Pescatello, 2010).

In contrast, to the increase in blood flow and SBP that occurs during exercise, total peripheral resistance normally falls. This is in response to the action of the resistance vessels in the active muscles that dilate to promote an increase in blood flow to meet the elevated metabolic demands of this tissue. This decreased resistance allows for a more rapid movement of blood out of the arteries and through the capillary system in the muscles during diastole of the heart. As a result, arterial pressure falls more quickly during diastole, and, consequently, the DBP normally falls or remains unchanged from resting values, even during intense rhythmic exercise. In cases of hypertension or hardening of the arteries, the normal dilation of the vessels with exercise may be inhibited, and DBP may rise considerably. It is recommended that exercise be stopped if DBP becomes elevated above 115 mmHg (Gibbons et al., 1997; Thompson, Gordon, and Pescatello, 2010).

Many different factors can influence resting and exercise blood pressure. In general, these influences can be grouped into physiologic/internal and environmental/external factors. As already discussed, cardiac output, total peripheral resistance, the condition or elasticity of the blood vessels, and blood viscosity are physiologic/internal factors that can affect blood pressure. In addition, since the circulatory system is a closed system, a change in blood volume can affect blood pressure. For example, an increase in blood volume as a result fluid retention by the kidneys can cause blood pressure to increase. This fact is the basis for some treatments for hypertension in which medications are given to cause a decrease of blood volume, and this decrease, in turn, results in a fall in blood pressure. Some environmental/external factors that also affect blood pressure are body position, exercise, temperature, altitude, emotions, food, and drugs. Thus, when measuring blood pressure, especially when evaluating an individual for suspected hypertension, care should be taken to standardize the environmental conditions under which blood pressure is measured.

The physical condition in which resting SBP or DBP is consistently elevated above normal levels is termed hypertension. It has been estimated that the number of adults ≥ 20 years of age who suffer from hypertension may be as high as 74.5 million, nearly 34% of the adult population of the United States, and the prevalence is nearly equal between men and women (Lloyd-Jones et al., 2010). The Joint National Committee on Detection, Evaluation, and Treatment of High Blood Pressure has published standards for classifying and treating elevated blood pressures (Table 7.1) (Chobanian et al., 2003).

TABLE 7.1 — Recommended standards for classifying normal and hypertensive blood pressures in adults 18 years of age and older

BP Classification	Systolic BP (mmHg)		Diastolic BP (mmHg)	Lifestyle Modification	Medication for BP
Normal	<120	and	<80	Encourage	No
Prehypertension	120–139	or	80–89	Yes	Yes
Stage 1 hypertension	140–159	or	90–99	Yes	Yes
Stage 2 hypertension	≥ 160	or	≥ 100	Yes	Yes

Adapted from The Seventh Report of the Joint National Committee on Prevention, Detection, Evaluation, and Treatment of High Blood Pressure; Chobanian et al., 2003.

It is well established that hypertension is a powerful risk factor for death from coronary heart disease and stroke, and that the risk increases continuously from lowest to highest blood pressure values, and is independent of other risk markers (Chobanian et al., 2003; Rutan et al., 1988). Hypertension is also associated with such medical problems as congestive heart failure, damage to the kidneys, aortic dissection, and peripheral vascular disease (Chobanian et al., 2003). It is encouraging to note that reducing blood pressure in those with hypertension can lower their chances of developing heart problems or suffering stroke (Chobanian et al., 2003), and that in some people regular exercise may help reduce high blood pressure (Hagberg, Park, and Brown, 2000). Thus, from a public health perspective it is very important to identify those with persistent hypertension. Appropriate therapy can then be instituted and the deleterious effects of hypertension can be prevented or reduced in severity.

Hypotension may be defined as abnormally low blood pressure. It is usually an indicator of insult or injury to the body, such as caused by shock, myocardial infarction, or drugs. These conditions are seldom encountered in an exercise setting. However, hypotension can present a problem for exercising individuals, especially after very vigorous exercise. During upright, dynamic exercise, such as running, the rhythmic contraction of the skeletal muscles aids in pumping blood back to the heart. As discussed previously, during high-intensity exercise the vascular beds of the exercising muscles dilate, lowering resistance and allowing for an increase in blood flow. If exercise is suddenly interrupted, as for example at the end of a maximal treadmill stress test, and the person tested is allowed to stand still, blood may pool in the dilated vessels of the legs resulting in a series of events that leads to a decreased cardiac output, a precipitous fall in blood pressure, dizziness, and ultimately fainting. Similar symptoms can be experienced by standing very quickly after lying down. To prevent this from happening after intense exercise, such as after a maximal treadmill stress test, the cool-down period should include an active component, such as walking, or the subject should be allowed to lie down.

The measurement of blood pressure is common in clinical settings and is a routine part of most exercise testing procedures. Direct measurement is seldom employed, since this requires the placement of a needle and catheter in an artery and an external pressure transducer or manometer. Several indirect methods of measuring blood pressure have been developed, including ultrasound and various electronic assist methods to detect arterial pressure in the finger, wrist, and arm (Pickering et al., 2004). The most commonly used indirect method employs an inflatable pressure cuff, a mercury column, aneroid, or electronic pressure gauge, and a common stethoscope as shown in Figure 7.1. The cuff and pressure gauge are

FIGURE 7.1 — Sphygmomanometer and stethoscope in position on subject's arm to take resting supine blood pressure.

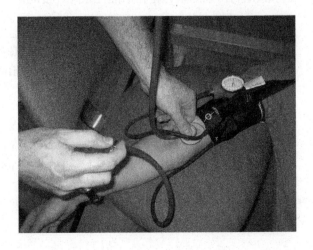

together referred to as a sphygmomanometer, and the method is referred to as indirect (or auscultatory) cuff sphygmomanometry.

The auscultatory procedures for measuring blood pressure by sphygmomanometry are really quite simple (Pickering et al., 2004). The cuff of the sphygmomanometer is inflated around a limb (usually the upper arm) until the pressure in the cuff exceeds that expected for SBP. At this cuff pressure, the underlying artery is collapsed preventing blood flow. A stethoscope is then positioned on the skin just above the artery, but distal to the cuff. Since the upper arm is commonly used for blood pressure measures, the stethoscope is placed over the brachial artery in the antecubital fossa at the elbow (see Figure 7.1). As the cuff is slowly deflated by opening its valve and the pressure fall is read off the manometer, a cuff pressure will be reached at which the SBP will exceed the pressure exerted by the cuff, blood will be forced through the compressed artery for a brief moment, and then the artery will snap closed as the arterial pressure falls during diastole. This brief opening and closing of the artery, along with the turbulent blood flow through the compressed artery, produces sounds that can be heard through the stethoscope. These changing sounds as pressure in the cuff is slowly released are called Korotkoff sounds after their discoverer. The blood pressure phases and sounds described by Korotkoff are shown in Table 7.2. The first Korotkoff sound marks SBP. As the volume of blood passing through the constricted artery increases with the continued release of cuff pressure, the sounds become progressively louder through phases II to III. The sounds then become suddenly muffled in phase IV as arterial constriction is reduced still further and diastolic pressure is approached. Finally the Korotkoff sounds disappear in phase V, which is usually taken as the diastolic blood pressure. Not all sounds are heard in every case.

The standard notation for recording blood pressure values is to write the SBP over the DBP, for example 120/80. However, in some individuals, particularly those who are well trained, phase V may occur at very low values after intense exercise, even as low as 0 mmHg. In these instances, phase IV is much closer to true diastolic pressure than is phase V, and phase IV should be recorded as the DBP.

TABLE 7.2 — Description of the blood pressure phases and Korotkoff sounds

Phases	Description of Sounds
Phase I	First appearance of clear tapping sound **Marks SBP**
Phase II	Murmuring or swishing sounds
Phase III	Crisper sounds increasing in intensity
Phase IV	Sounds suddenly become muffled
Phase V	Sounds disappear **Marks DBP**

Pickering et al., 2004.

Equipment Needed

Sphygmomanometer (mercury column or aneroid)

Stethoscope (single and teaching with two headsets)

Stationary cycle ergometer (mechanically or electronically braked)

Motorized treadmill

Metronome

TECHNIQUES AND PROCEDURES

The technique for measuring resting blood pressure described below is adapted from the recommendations of the American Heart Association for measuring blood pressure in humans (Myers et al., 2009; Pickering et al., 2004). Procedures for exercise blood pressure are adapted from the recommendations of the American Heart Association (Myers et al., 2009) and American College of Sports Medicine (Thompson, Gordon, and Pescatello, 2010).

GENERAL PROCEDURES FOR RESTING BLOOD PRESSURE MEASUREMENTS

1. Be sure to become familiar with the equipment to be used in this laboratory exercise before attempting to measure blood pressure on the subject. Procedures for use of the cycle and treadmill ergometers are thoroughly described in the Laboratory 1 entitled "Ergometry: The Measurement of Work and Power."
2. This laboratory exercise is most efficiently conducted in teams of three, one person to serve as the subject, one to perform the duties of the blood pressure technician, and one to control the ergometers and record data.
3. If required by the institution, ask the subject to read and sign an informed consent. See Laboratory 8 for an example of an informed consent which could be adapted for this blood pressure laboratory.
4. The following general guidelines should be observed to standardize the environment where blood pressure will be measured. Body position, exercise, temperature, altitude, emotions, food, and drugs affect blood pressure, so these factors should be controlled when standardizing blood pressure procedures.
   a. The physical setting should be in a comfortably warm room, clean, neat, quiet, and free from distracting activities.
   b. Greet the subject in a calm and reassuring manner. Ask the subject about his or her most recent use of beverages containing caffeine, nicotine products, drugs including alcohol, prescribed medications, and any recent exercise. For accurate blood pressure measurements, the subject should not be under the influence of adrenergic stimulants or alcohol. If the subject is taking antihypertensive medication, the time since the last dose was taken should be noted. Also ask the subject what his or her blood pressure usually measures. Record this information on the Blood Pressure Worksheet found at the end of this laboratory.
   c. The subject should be dressed with unrestrictive clothing that does not cover or interfere with the cuff placement on the arm or with arm blood flow.
5. In clinical practice, resting blood pressure measurements are usually taken with the subject comfortably seated or lying down for 5 minutes prior to measurement, legs uncrossed, back and arm supported with the arm at heart level (midpoint of the sternum, at fourth intercostal space). Arm position is very important since blood pressure can vary as much as 2 mmHg for every inch the arm is above or below heart level (Pickering et al., 2004). In this laboratory, each student will measure resting blood pressure with the subject in the supine, sitting, and standing positions, starting with the subject in the supine position.
6. Locate by palpation the strongest brachial artery pulse in the antecubital fossa at the bend of the elbow in both arms. When learning to measure blood pressure and when subsequent blood pressure

measurements will be made during exercise, mark this location with a small "x" in ink so that it can be readily located quickly and accurately during repeated measurements.

7. Wrap and secure the deflated cuff about 2.5 cm above the antecubital space of either arm so that the inflatable bladder in the cuff is centered on-line with the marked brachial artery. Most cuffs have a mark to aid in this correct placement (refer to Figure 7.1).
    a. Check to ensure that the cuff size is appropriate for the individual. Use of an improper cuff size may lead to errors in blood pressure measurement. For example, use of an undersized cuff will result in a falsely elevated blood pressure value. The air bladder length in the cuff should be 80% and the width at least 40% of the arm circumference. Recommended sizes for adults are:
        i. 12 × 22 cm for arm circumferences 22–26 cm (small adult size)
        ii. 16 × 30 cm for arm circumferences 27–34 cm (adult size)
        iii. 18 × 36 cm for arm circumferences 35–44 cm (large adult size)
        iv. For arm size ≥ 45 cm, use an adult thigh size at least 42 cm in length
        v. Smaller sizes are available and should be used for children.
8. Be sure the arm is supported at the elbow and held fully extended at heart level (level with midpoint of the sternum, fourth intercostal space) during the measurement of blood pressure. It is advisable for the technician to support the arm on a table or chair, or use one hand under the subject's extended elbow to provide lift and support and to keep it at the correct heart level. This will be particularly important during exercise.
9. The technician should now insert the earpieces of the stethoscope into the ears, hold the subject's elbow fully extended level with the heart, and apply the bell of the stethoscope with gentle pressure to the center of the "x" previously marked to locate the brachial artery.
    a. At this step it is very important for the inexperienced technician to use one side of a dual-headset teaching stethoscope and another technician or instructor experienced in blood pressure measurements listening on the second headset. The experienced technician/instructor can help the student identify the Korotkoff sounds and verify the blood pressure readings of the student, greatly aiding in accurate mastery of the technique.
    b. Check to be certain that no air space exists between the stethoscope and the skin, and that the stethoscope is not touching the subject's clothing, the cuff, or any of the cuff tubing.
10. Hold the bell of the stethoscope on the brachial artery with one hand, and with the other hand turn the bladder valve clockwise (to the right) to carefully close it (do not over-tighten), and inflate the cuff to a reading about 30 mmHg above the expected systolic pressure (usually to about 140–160 mmHg at rest).
    a. Once the cuff has been inflated, *slowly* turn the bladder valve counter-clockwise (to the left) to open it and release the pressure *gradually* at a rate of about 2 to 3 mmHg/second.
    b. Closely observe the pressure fall on the manometer and listen intently for the Korotkoff sounds to identify sounds I and V corresponding to SBP and DBP, respectively (see Table 7.2). Repeat as often as necessary until (1) the Korotkoff sounds for SBP and DBP are unmistakably heard; (2) the blood pressure measures are validated by an experienced technician/instructor listening on the teaching stethoscope; and (3) the measures are repeatable.
        i. **Note:** The cuff pressure must be fully released for about 10–15 seconds between measures to allow blood to recirculate through the forearm.
    c. Once accuracy is assured, record the SBP and DBP on the Blood Pressure Worksheet.

11. Take two readings in each arm separated by as much time as is practical. If the two readings in any arm differ by more than 5 mmHg, additional measurements should be made to verify the measured values (Pickering et al., 2004). If several measurements are required to achieve reproducibility, the average of the last two should be recorded as the actual blood pressure.
    a. Occasionally a subject may have a higher blood pressure measured in one arm compared with the other. If this is the case, the blood pressure from the arm with the highest reading should be recorded as the actual blood pressure. All subsequent blood pressure measurements taken during exercise should be performed in this arm.
    b. Blood pressure readings from manual sphygmomanometers are usually given in even numbers, e.g., 126/76. Electronic instruments may give odd numbers and should be used if available.
12. After the supine measurements have been made, repeat the resting procedures with the subject seated and then standing. Record the resting values on the Blood Pressure Worksheet.

MEASUREMENT OF EXERCISE BLOOD PRESSURE

The skills required to measure exercise blood pressure are not technically difficult but require a substantial amount of practice to be mastered. Since the exercise technician will frequently encounter the need to measure blood pressure on subjects exercising on the stationary cycle ergometer and the motorized treadmill, both will be used in this laboratory exercise. It is helpful when first learning the technique to practice with the subject on the cycle ergometer. This exercise apparatus is usually quieter to operate, and there is less upper body movement of the subject during exercise than on the treadmill, making it easier to measure blood pressures. The general techniques for measuring resting blood pressure apply to exercise, with the following exceptions and additions:

1. It is clearly impractical to perform duplicate measures in each arm during exercise as is done at rest. Therefore, whenever possible the arm in which the highest resting blood pressures were recorded should be used exclusively during exercise.
2. As with resting blood pressures, it is very important that novice technicians use a teaching stethoscope with an experienced technician/instructor while learning the technique.
3. Measure blood pressure once at each stage, as closely as is practical to the end of the stage, and after steady-state has been achieved following the increase in workload. Allow between 30 and 60 seconds for the blood pressure measurement. Start the measurement about 1 minute prior to the end of each exercise stage. If difficulty is encountered, or the technician is unsure of the reading, the measurement should be quickly repeated.

BLOOD PRESSURE MEASUREMENT ON THE STATIONARY CYCLE ERGOMETER

This part of the laboratory exercise should be completed after all resting data has been collected and recorded. Refer to Figure 7.2 for the position of the subject and technician to take blood pressure on the cycle ergometer.

1. Check to be sure the stationary cycle ergometer has been recently calibrated. If calibration is in question, follow the calibration procedure given in the Appendix before completing this laboratory exercise.
2. Ask the subject to be seated on the stationary cycle ergometer. Adjust the seat so there is a slight bend (15–25°) in the subject's legs when they are extended at the bottom of the pedal stroke.

FIGURE 7.2 — Subject and technician positioning for taking blood pressures on the cycle ergometer.

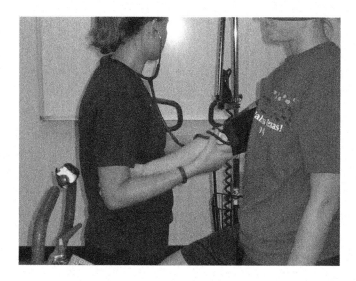

3. The subject will be asked to pedal at a frequency of 50 revolutions-per-minute (rpm). Instruct the subject to maintain this pedal cadence by viewing the tachometer on the cycle ergometer, if so equipped and accuracy has been calibrated.

   a. On models without a tachometer, set a metronome so that it emits a signal (light, sound, or both depending on the model) 100 times per minute. At this metronome rate, the subject should be instructed to push first with one foot then the other in time with the metronome signal. This will ensure that the subject completes one pedal revolution every two metronome signals, which is equivalent to a rate of 50 rpms.

   b. It is *very important* that the prescribed rpms be maintained throughout the exercise, since the work rate on mechanically braked cycle ergometers (such as a Monark®) is dependent upon the pedal frequency (review Laboratory 1 for cycle ergometer operational instructions). Note that the workload on most electronically braked cycle ergometers (e.g., Lode®) is *independent* of pedal rpms, since these ergometers automatically adjust the resistance during pedaling to maintain a prescribed power setting, usually in watts. Therefore, on most electronically braked ergometers, the subject can pedal at any self-selected pedal frequency.

4. With the subject seated on the ergometer, measure the pre-exercise blood pressure following the procedures above for resting blood pressure. Record the accurate value on the Blood Pressure Worksheet.

5. Secure the cuff to the upper arm so that it will not slip down during exercise. Tape the cuff to the shoulder if necessary.

6. **DECISION POINT:** Pre-exercise blood pressure values greater than 200 mmHg systolic or 110 mmHg diastolic are considered relative contraindications for exercise (Thompson, Gordon, and Pescatello, 2010). Do not continue with the exercise portion of this laboratory if the subject displays either of these blood pressure values.

7. Instruct the subject to begin pedaling at 50 rpms and set the resistance to 0.25 kp (75 kpm · min^{-1} or 12.5 watts). Continue at this warm-up load until the subject is able to accurately maintain the pedal cadence, usually about 1 minute.

8. Advance to stage 1 of the exercise protocol as shown on the Blood Pressure Data Worksheet, and continue through recovery. Note that each stage, including recovery, is 2 minutes in duration.

9. **STOP THE EXERCISE** if at any time: (1) the subject requests to stop; (2) they are unable to maintain the cadence; or (3) their systolic blood pressure exceeds 250 mmHg or diastolic 115 mmHg (Thompson, Gordon, and Pescatello, 2010).
10. Begin measuring blood pressure 1 minute before the end of each stage, or earlier if needed, to ensure that the blood pressure measurement can be completed before the load increase is made for the next stage. Follow these steps for accurate blood pressure measures (refer to Figure 7.2).
    a. Ask the subject to release the handlebars with the arm that has the cuff attached.
    b. Grasp the subject's arm under the elbow with one hand, and gently lift it until fully extended and positioned at about heart level. To further steady the arm once free from the handlebars, the subject can be asked to grasp the inside upper portion of the arm the technician is using to steady the subject's elbow.
    c. Place the stethoscope over the small "x" used to mark the brachial artery at rest, and measure the exercise blood pressure following the procedures given above for resting blood pressure. Again be sure the blood pressure measures are validated by an experienced technician/instructor using a teaching stethoscope with the student.
    d. In learning the technique, the novice student may require an extended period of time to accurately compete the procedure, or may require an additional attempt to obtain an accurate reading. In this case, do not advance to the next stage of exercise until the blood pressure measures have been completed.
    e. Be sure to fully deflate the bladder between blood pressure measurement attempts to ensure normal blood flow through the lower arm below the cuff.
    f. Record the blood pressure values on the Blood Pressure Worksheet, and continue the protocol as prescribed on this form recording blood pressures at each stage and recovery.

BLOOD PRESSURE MEASUREMENT ON THE TREADMILL

1. After sufficient time of rest recovery from the cycle ergometer exercise, escort the subject to the treadmill. If the subject is unfamiliar with treadmill exercise, the technician should demonstrate the safe and proper manner to initiate walking on the treadmill. These procedures were fully described in Laboratory 6 *Resting and Exercise Electrocardiography*. Refer to Figure 7.3 for the positioning of the subject and technician for measuring blood pressure on the treadmill.
2. Be sure the treadmill belt is not moving. Instruct the subject to step onto the stationary treadmill belt and stand quietly.
    a. Measure the standing, pre-exercise blood pressure at this time following the procedures given previously in this laboratory for the measurement of resting blood pressure.
    b. Record the resting pre-exercise value on the Blood Pressure Worksheet.
    c. Secure the cuff to the upper arm so that it does not slip down during exercise; tape in place if necessary.
3. **DECISION POINT:** If the resting SBP exceeds 200 mmHg or the DBP exceeds 110 mmHg, **exercise should not be performed at this time** (Thompson, Gordon, and Pescatello, 2010).
4. Ask the subject to straddle the belt, and then start the treadmill at 1.7 mph and 0% grade. Ask the subject to hold the treadmill rails while stepping onto the moving treadmill belt. As soon as they

FIGURE 7.3 — Subject and technician positioning for taking blood pressures on the treadmill.

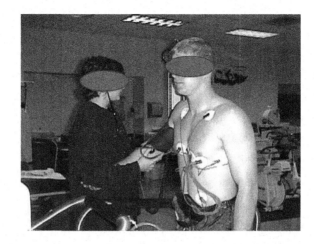

are walking comfortably, ask him or her to release the handrails and walk normally. Advance the treadmill to stage 1 shown on the Blood Pressure Worksheet, and continue the protocol through recovery.

5. **STOP THE EXERCISE** if at any time (1) the subject requests to stop, (2) the subject is unable to maintain the pace, or 3) the subject's blood pressure exceeds 250 mmHg systolic or 115 mmHg diastolic (Thompson, Gordon, and Pescatello, 2010).
6. Using the same basic technique as was described for measuring blood pressure on the cycle ergometer, repeat the procedures for each treadmill stage, including recovery. Care must be taken to limit the movement of the arm and to ensure that it is fully extended when taking blood pressure measurements. Begin each measurement at least 1 minute before the end of the stage, and repeat as necessary. Do not advance to the next stage of exercise until an accurate blood pressure measure is completed.
7. Record the accurate blood pressure values appropriately on the Blood Pressure Worksheet.
8. At the end of recovery, inform the subject of your intention and then stop the treadmill belt.
9. Remove the blood pressure cuff at the completion of all procedures and dismiss the subject.

QUESTIONS AND LEARNING ACTIVITIES

1. Using a computer program of choice, construct a line chart of the SBP, DBP, PP, and MBP values measured during rest, stage 1 and 2 of exercise, and recovery on the cycle ergometer. Construct a second line chart of the same variables obtained on the treadmill.
2. Visually review the charts constructed in activity 1, and describe how these blood pressure variables changed as the exercise intensity increased. Provide physiologic rationale for the changes in blood pressure measured with exercise and then during recovery.
3. Was there a difference in the blood pressure responses on the stationary cycle ergometer compared to the treadmill? Provide physiologic reasons why or why not.
4. Why is it necessary to monitor exercise blood pressure so closely in persons suspected of having coronary heart disease or hypertension?

QUESTIONS AND LEARNING ACTIVITIES FOR GOING FURTHER

1. Explain how the following conditions might affect the blood pressure response to exercise.
   a. Being well-conditioned.
   b. Having coronary heart disease.
   c. Being poorly conditioned but free from disease.

2. A walking cool-down period is usually recommended after a maximal exercise test on a treadmill. What hemodynamic consequence could be encountered if a subject were permitted to stand upright without moving immediately after heavy exercise? Explain this consequence using physiologic rationale.

REFERENCES

Chobanian, A. V., G. L. Bakris, H. R. Black, W. C. Cushman, L. A. Green, J. L. Izzo, D. W. Jones, B. J. Materson, S. Oparil, J. T. Wright, and E. J. Roccella. "The Seventh Report of the Joint National Committee on Prevention, Detection, Evaluation, and Treatment of High Blood Pressure—The JNC 7 Report." *JAMA* 289 (2003): 19, 2560–72.

Gibbons R. J., G. J. Balady, J. W. Beasley, J. T. Bricker, W. F. C. Duvernoy, V. F. Froelicher, D. B. Mark, T. H. Marwick, B. D. McCallister, P. D. Thompson, W. L. Winters, and F. G. Yanowitz. ACC/AHA Guidelines for Exercise Testing—A Report of the American College of Cardiology American Heart Association Task Force on Practice Guidelines (committee on exercise testing). *J. Am. Coll. Cardiol.* 30 (1997): 1, 260–311.

Hagberg, J. M., J. J. Park, and M. D. Brown. "The Role of Exercise Training in the Treatment of Hypertension: An Update." *Sports Med.* 30 (2000): 3, 193–206.

Lloyd-Jones D., R. J. Adams, T. M. Brown, M. Carnethon, S. Dai, G. De Simone, T. B. Ferguson, E. Ford, K. Furie, C. Gillespie, A. Go, K. Greenlund, N. Haase, S. Hailpern, P. M. Ho, V. Howard, B. Kissela, S. Kittner, D. Lackland, L. Lisabeth, A. Marelli, M. M. McDermott, J. Meigs, D. Mozaffarian, M. Mussolino, G. Nichol, V. L. Roger, W. Rosamond, R. Sacco, P Sorlie, R. Stafford, T. Thom, S. Wasserthiel-Smoller, N. D. Wong, and J. Wylie-Rosett. American Heart Association Stat Comm Stroke. Heart Disease and Stroke Statistics—2010 Update: A Report from the American Heart Association. *Circulation* 121 (2010): 7, E46–E215.

Myers J., R. Arena, B. Franklin, I. Pina, W. E. Kraus, K. McInnis, and G. J. Balady. Recommendations for Clinical Exercise Laboratories: A Scientific Statement from the American Heart Association. *Circulation* 119 (2009): 24, 3144–61.

Pickering T., J. Hall, L. Appel, B. Falkner, J. Graves, and M. Hill. Recommendations for Blood Pressure Measurement in Humans and Experimental Animals—Part 1: Blood Pressure Measurement in Humans—A statement for Professionals from the Subcommittee of Professional and Public Education of the American Heart Association Council on High Blood Pressure Research. *Hypertension* 45 (2005): 1, 142–161.

Rutan, G. H., L. H. Kuller, J. D. Neaton, D. N. Wentworth, R. H. McDonald, and W. M. Smith. Mortality Associated with Diastolic Hypertension and Isolated Systolic Hypertension among Men Screened for the Multiple Risk Factor Intervention Trial. *Circulation* 77 (1988): 3, 504–14.

Thompson, W. R., N. F. Gordon, and L. S. Pescatello (eds.). *ACSM's Guidelines for Exercise Testing and Prescription,* Philadelphia: Lippincott Williams & Wilkins, 2010.

BLOOD PRESSURE WORKSHEET

Subject Name: Jason   Age: 25   Gender: male   Date: _____

Medical History: none

Medications/Doses: none   Last Taken: _____

Past 24 hr: Caffeine no   Nicotine no   Drugs/Alcohol no

Other: _____

Last Exercise: 24 hrs   Usual BP (if known): 120/80   BP Tech Name: _____

	RESTING DATA			
	Trial 1		Trial 2	
Body Position	Right Arm	Left Arm	Right Arm	Left Arm
Supine Blood Pressure (mmHg)	116/62		118/62	
Sitting Blood Pressure (mmHg)	118/72		118/72	
Standing Blood Pressure (mmHg)	122/78	128/78	122/80	130/76

		STATIONARY CYCLE ERGOMETER DATA				
Stage	Time	RPM	kp	kpm (Watts)	BP	Comments
Pre-Ex	Seated	-	-	-		
1	0–2	50	.5	150 (25)		
2	2–4	50	1.0	300 (50)		
3	4–6	50	1.5	450 (75)		
Recovery	6–8	50	.5	150 (25)		

		TREADMILL DATA			
Stage	Time	Speed (mph)	Grade (%)	BP	Comments
Pre-Ex	Standing	-	-	124/76	
1	0–2	2.0	10	130/70	
2	2–4	2.5	12	132/72	
3	4–6	3.0	14	136/78	
Recovery	6–8	2.0	0	132/76	

CHAPTER 8

SUBMAXIMAL EXERCISE TESTING FOR ESTIMATING AEROBIC CAPACITY

OUTCOME OBJECTIVES

Following successful completion of this exercise, the student should be able to:

1. Conduct a submaximal exercise test using a treadmill, cycle ergometer, and bench-step protocol.
2. Calculate estimated maximal oxygen uptake ($\dot{V}O_{2max}$) using the data collected during the submaximal exercise test.
3. Measure heart rate by palpation during submaximal exercise testing.
4. Describe the sources of error and benefits of estimating $\dot{V}O_{2max}$ using submaximal procedures.

INTRODUCTION

Cardiorespiratory endurance may be defined as the ability to perform dynamic muscular work at moderate to high levels of intensity for prolonged periods of time. Many other terms are found in exercise physiology literature that have nearly the same meaning, such as aerobic capacity, cardiorespiratory fitness, endurance capacity, and functional aerobic capacity. Cardiorespiratory endurance depends, to a large extent, on the ability of the respiratory, cardiovascular, and skeletal muscle systems working together to take in oxygen from the atmosphere, transport it to the muscles, and use it in aerobic metabolism. Generally speaking, the greater the ability of these three systems to perform their roles in the transport and utilization of oxygen, the greater the amount of work that can be performed without undue fatigue. It follows that a person who has a comparatively greater capacity to perform work has by definition a higher cardiorespiratory endurance.

$\dot{V}O_{2max}$ is an index of the *maximal* functional capacity of these three systems—cardiovascular, respiratory, and muscular—working together. It can be defined as the greatest amount of oxygen that an individual can take up from the atmosphere, transport, and utilize during strenuous exercise to voluntary exhaustion, and it has become the accepted criterion measure of cardiorespiratory endurance (Thompson, Gordon, and Pescatello, 2010). Measured $\dot{V}O_{2max}$ is often expressed as a volume of oxygen consumed per minute ($L \cdot min^{-1}$), but when used to compare the cardiorespiratory endurance capacity of one individual to another, it is usually expressed relative to body weight as ml O_2 per kg body weight per min ($ml \cdot kg^{-1} \cdot min^{-1}$). Average values for $\dot{V}O_{2max}$ may range from a low of about 35 $ml \cdot kg^{-1} \cdot min^{-1}$ in 40 to 50-year-old, sedentary men to a high of over 75 $ml \cdot kg^{-1} \cdot min^{-1}$ in young endurance runners; values for women are typically 10% to 20% lower. Acceptable norm values for adults by age and gender are given in Table 8.1.

TABLE 8.1 — $\dot{V}O_{2max}$ norm values in ml · kg^{-1} · min^{-1} by age and gender

Rating Category	Age 18–25 M	18–25 F	26–35 M	26–35 F	36–45 M	36–45 F	46–55 M	46–55 F	56–65 M	56–65 F	>65 M	>65 F
Excellent	≥63	≥58	≥58	≥54	≥53	≥46	≥47	≥42	≥43	≥38	≥38	≥33
Good	53–62	48–57	50–57	46–53	44–52	39–45	40–46	35–41	37–42	32–37	33–37	28–32
Above Average	47–51	42–47	44–49	40–45	40–43	34–38	35–39	31–34	33–36	28–31	29–32	25–27
Average	43–46	39–41	40–43	35–39	35–39	31–33	32–34	28–30	30–32	25–27	25–28	22–24
Below Average	38–42	34–38	35–39	31–34	32–34	28–30	29–31	25–27	26–29	22–24	22–24	20–21
Poor	31–37	29–33	31–34	26–30	27–31	23–27	26–28	21–24	22–25	19–21	20–21	17–19
Very Poor	≤30	≤28	≤30	≤25	≤26	≤22	≤25	≤19	≤21	≤18	≤19	≤16

Modified from Golding, Myers, and Sinning, 1989.

It is most common to measure $\dot{V}O_{2max}$ directly when accuracy and precision are required, such as in research laboratories, testing elite athletes, and in assessing the cardiopulmonary status of cardiac or pulmonary patients. In modern laboratories, this is usually accomplished by measuring the volume and the O_2 and CO_2 concentrations of inhaled or exhaled air with automated, computerized equipment during a symptom-limited maximal graded exercise test on a treadmill or cycle ergometer.

While direct measurement of $\dot{V}O_{2max}$ is ideal, the application of these procedures to mass testing and for assessment of general physical fitness is often not practical. The equipment to measure $\dot{V}O_{2max}$ is expensive and requires a trained technician to operate, the procedures are quite time consuming, and the subject must be highly motivated to reach maximal exertion. To add to the staffing and cost requirements, it is generally recommended that maximal exercise testing of individuals with two or more cardiovascular disease risk factors or who have documented cardiac, pulmonary, or metabolic disease be directly supervised by a physician or conducted by trained health care professionals with a physician in the immediate area (Myers et al., 2009; Thompson, Gordon, and Pescatello, 2010). In contrast, physician supervision is not required when submaximal testing is performed on men and women of any age and risk status, as long as they do not have symptoms or disease. Therefore, submaximal assessment of cardiorespiratory endurance is often employed in health clubs, corporate fitness programs, and in other situations where mass testing is required or where equipment and personnel with advanced training are not available. Owing to the demand and popularity of submaximal testing, a number of procedures for estimating $\dot{V}O_{2max}$ from submaximal exercise have been developed. General advantages and disadvantages or submaximal testing are shown in Table 8.2.

TABLE 8.2 — General advantages and disadvantages of submaximal exercise testing

Advantages
1. Tests are relatively inexpensive and do not require the use of costly equipment
2. Testing personnel require little training to obtain reliable results
3. Some submaximal tests allow for mass testing
4. Test duration is usually shorter than direct measurement tests
5. Tests do not require that a maximal work effort be given by the subject which increases the safety margin of the test, especially in high risk subjects
6. No physician supervision necessary unless subject has symptoms or disease
7. $\dot{V}O_{2max}$ can be estimated
8. Heart rate and blood pressure can be monitored for appropriate responses to exercise at intensities usually encountered during training
9. Useful to qualify changes in cardiorespiratory endurance with training
Disadvantages
1. Maximal heart rate, blood pressure, and rate-pressure-product not measured
2. $\dot{V}O_{2max}$ is not measured directly - errors in prediction range from about 10 to 20%
3. Limited diagnostic usefulness since <85% maximal heart rate achieved by subject and ECG frequently not recorded
4. True, measured maximal heart rate is unavailable to use in prescribing exercise

FIGURE 8.1 — Relationship between heart rate and $\dot{V}O_2$.

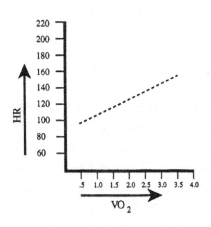

Predictions of $\dot{V}O_{2max}$ from submaximal exercise data are often based on the assumption that a linear relationship exists between $\dot{V}O_2$, workload, and heart rate (Figure 8.1). This is generally true for workloads above the intensity at which stroke volume becomes maximal, which occurs on average at an exercise intensity of about 40% of $\dot{V}O_{2max}$ (Ekblom et al., 1968). In a 20-year-old individual, this would correspond to a heart rate of about 120 bpm. Since cardiac output equals the product of heart rate and stroke volume, it follows that above this exercise intensity the major increase in cardiac output, and consequently in $\dot{V}O_2$, is directly related to an increase in heart rate. On the other hand, relatively high exercise intensities should be avoided when doing submaximal testing, since the relationship between heart rate and $\dot{V}O_2$ becomes curvilinear near maximum exercise. For this reason it is generally recommended that submaximal test should be conducted at an intensity which will elicit a steady-state heart rate between about 115 and 150 bpm (Åstrand and Åstrand, 2003).

Once maximum heart rate and the relationship between heart rate and $\dot{V}O_2$ or workload are known for an individual, $\dot{V}O_{2max}$ can be predicted by calculating what the $\dot{V}O_2$ would be at maximum heart rate. If a true, *measured* maximum heart rate achieved on a maximal stress test is not available, it can be reasonably estimated using the following equations (Gellish et al., 2007; Thompson, Gordon, and Pescatello, 2010).

$$\text{Predicted Maximal Heart Rate} = 207 - (0.7 \times \text{age in years})$$

$$\underline{OR} = 220 - \text{age in years}$$

It should be noted that the practice of estimating maximum heart rate from age contributes to the error in predicting $\dot{V}O_{2max}$, since *measured* maximum heart rates can vary as much as ± 15 bpm for individuals of the same age (Thompson, Gordon, and Pescatello, 2010). The general concept for estimating $\dot{V}O_{2max}$ from submaximal exercise and heart rate data is illustrated in Figure 8.2.

Numerous modalities and procedures have been used to provide the exercise stimulus for submaximal exercise testing. Valid procedures include the use of treadmill or track walking/running, stationary cycle ergometry, and stair-stepping. Mass testing procedures often make use of endurance runs (Cooper, 1968), walks (Rockport Walking Institute, 1986), or bench-stepping (Brouha, 1943; McArdle et al., 1972). In these cases, cardiovascular capacity or $\dot{V}O_{2max}$ is predicted from distance or time completed on a measured track, or from recovery heart rates measured immediately after exercise. Mass testing procedures are employed in situations that call for a large number of individuals to be tested in a short period of time; for example, in the testing of military recruits or students enrolled in physical fitness classes in school settings. The accuracy of these procedures varies considerably depending on several characteristics of the subjects tested, such as age, gender, body fat, and running or walking efficiency (Heyward, 1998). Moreover, the fact that many mass testing procedures do not readily lend themselves to the measurement of physiologic responses during exercise (e.g., blood pressure and heart rate), their usefulness is seriously limited for identifying individuals with abnormalities associated with exercise. For example, cardiac

FIGURE 8.2 — Illustration of the general concept for predicting $\dot{V}O_{2max}$ from submaximal heart rate.

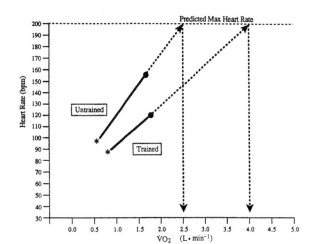

arrhythmias and abnormal blood pressure responses, which might occur during exercise, would likely be missed using mass testing procedures.

A bench-stepping exercise test is detailed in this laboratory as an example of a typical mass-testing procedure to estimate $\dot{V}O_{2max}$. In bench step tests, the exercise workload is imposed by stepping up and down on a bench of known height and at a specified step rate. The exercise is continued for a specific length of time, usually from three to five minutes. It is assumed that a steady-state heart rate will be reached during this work bout, and, as for other submaximal tests, that the heart rate will be proportional to the submaximal $\dot{V}O_2$. Another assumption underlying this test is the fact that the heart rate is higher during exercise and returns to normal more slowly after exercise in subjects who are relatively less fit. Thus, $\dot{V}O_{2max}$ is related to the recovery heart rate measured the first few seconds after stopping steady-state exercise; that is, the higher the $\dot{V}O_{2max}$, the more quickly heart rate returns to resting levels after exercise. This fact is the basis for predicting $\dot{V}O_{2max}$ from recovery heart rates measured shortly after the bench step exercise.

For more individualized and complete submaximal testing, the stationary cycle ergometer and the treadmill are the most commonly used devices. Both readily permit the measurement of blood pressure and heart rate during exercise. The electrocardiogram can be recorded to assess heart rate, heart rhythm, and ischemic changes. Heart rate monitor watches are marketed by many manufacturers and are a relatively inexpensive way to accurately measure heart rate during exercise. Simple palpation of the carotid or radial pulse is also often used to measure heart rates. Although palpation is of no use in detecting ischemic changes, when used by a trained technician it is a valid method of measuring heart rate, and, to a lesser extent, can be used to detect some types of arrhythmias, such as premature ventricular contractions. Because a cycle ergometer, compared to the treadmill, is generally cheaper to purchase and maintain, requires less space, is quieter to operate, and allows for greater ease of measuring physiologic variables, it is often the equipment of choice when conducting submaximal exercise testing for cardiovascular endurance.

In this laboratory, treadmill, cycle ergometer, and bench-stepping submaximal exercise tests will be used to predict $\dot{V}O_{2max}$.

Equipment Needed

Motorized treadmill

Stationary cycle ergometer (Monark®)

Step box 16.25" tall

Sphygmomanometer, blood pressure cuff, and stethoscope

Metronome

Stopwatch

Heart rate watch monitor

TECHNIQUES AND PROCEDURES

In contrast to the usual staff requirements of two trained persons to carry out the procedures for maximal graded exercise testing, one well-trained exercise technician can usually complete all submaximal testing procedures. However, this laboratory exercise is most efficiently conducted in teams of three: one person to serve as the subject, one to perform the required measurements, and one to control the ergometers, timing devices, and record data. Procedures for conducting submaximal exercise tests using a treadmill, a stationary cycle ergometer, and a step box are detailed in this laboratory.

GENERAL PROCEDURES TO COMPLETE BEFORE TESTING THE SUBJECT

1. Communicate to the subject at least one day in advance information about the general nature of the exercise test, that is, what to expect regarding the type of exercise and what types of measures will be taken.
2. Subjects should wear clothing that is comfortable and loose fitting and include walking shoes suitable for exercise. A T-shirt or sleeveless shirt is necessary so that blood pressures can be easily taken without clothing restrictions.
3. Instruct subjects to avoid heavy meals at least 2 hours before arriving at the lab and to report well-hydrated.
4. Become operationally familiar with all equipment that will be used in this laboratory experience before attempting to test subjects. Review the procedures for treadmill, cycle ergometer, and step testing in Laboratory 1 entitled "Ergometry: The Measurement of Work and Power." Completing this laboratory also requires experience in measuring exercise blood pressure. These techniques have been thoroughly described in Laboratory 7 "Measuring Resting and Exercise Blood Pressure."
5. Check to ensure the recent calibration of all equipment to be used for testing. Generally, the treadmill and stationary cycle ergometer should be calibrated at least once per month. Refer to the calibration procedures in the Appendix if calibration is needed.
6. On the day of testing, check laboratory environmental conditions to ensure that the temperature range is between 20°C and 22°C (68°F to 71.6°F) and the relative humidity less than 60%. It is important to standardize these conditions from day to day, especially if testing is to be repeated later on the same individual, since these environmental factors can alter the relationship between heart rate and $\dot{V}O_2$ on which predicting $\dot{V}O_{2max}$ is based (Thompson, Gordon, and Pescatello, 2010).

PRELIMINARY PROCEDURES

1. Upon arrival, and before any testing procedures are performed, ask the subject to read the "Informed Consent for Submaximal Exercise Testing." A model adapted from the ACSM Guidelines for Exercise Testing and Prescription (Thompson, Gordon, and Pescatello, 2010) can be found at the end of this laboratory chapter. (*Important note:* Check with the appropriate institutional authorities about the policy of the institution or business for any additional legal requirements for subject informed consent.)

   a. The testing personnel should encourage the subject to ask questions about the informed consent and the testing procedures.

   b. After all questions have been answered to the subject's satisfaction, ask the subject to sign and date the informed consent document. The testing personnel should then sign as witness. *No testing should be performed until this document is signed and witnessed.*

2. Ask the subject to complete the "Submaximal Testing Readiness Questionnaire" included in this laboratory to help identify any conditions that might limit the subject's exercise ability or make exercise unsafe. Review this questionnaire carefully.

   a. **DECISION POINT.** If any of the items are checked by the subject, the technician should verbally question the subject for clarification *before* starting any exercise testing procedures. Ask the subject to quantify high cholesterol or use of tobacco, if reported. *In the case of any checked items, no testing should be performed until approval to proceed is confirmed by consulting with a trained exercise or medical professional.*

   b. Before initiating exercise testing, the exercise technician must completely understand the potential effects on exercise of any medications listed by the subject. Differences in medication types or doses between tests performed several weeks or months apart may make test comparisons invalid. For example, a change in blood pressure medication from a beta-adrenergic blocker (e.g., propranolol) to an ACE-inhibitor between serial tests would likely result in a quite different heart rate response to submaximal exercise that is due to the change in medications, and not to a change in cardiorespiratory conditioning.

3. Rather than direct measurement using an electrocardiogram, heart rate during submaximal exercise testing is commonly measured by palpation of either the radial or carotid pulse, by auscultation of the heart with a stethoscope, or by using a heart rate watch monitor. In this laboratory, measurement of heart rates will be accomplished using the palpation technique, and verified by a heart rate watch monitor. The general method for measuring heart rate by palpating the pulse at either the radial or carotid artery follows here.

   a. Place the index and middle fingers together, then gently apply pressure with the fingertips to feel the pulse at the radial artery in the wrist or the carotid artery in the neck.

      i. <u>Carotid arterial pulse.</u> This pulse can be palpated in the neck on either side of the larynx. To locate, apply gentle pressure using the tips of the fingers to the point of the larynx, then slide the fingers slowly to either side until the pulse in located. Be careful to apply only *light* pressure to the carotid artery, since heavy pressure can stimulate the pressure-sensitive carotid baroreceptors causing a reflex slowing of the heart rate.

      ii. <u>Radial arterial pulse.</u> This pulse is located on the anterolateral aspect of the wrist in line with the base of the thumb. To palpate this pulse, place the fingertips on the anterior surface of the wrist at the base of the thumb. (Figure 8.3)

FIGURE 8.3 — Palpating the radial pulse to obtain heart rate.

   b. Start the stopwatch simultaneously with the pulse, and count the first beat as zero. The count can be continued for any designated period of time up to one minute. In practice, 6-, 10-, 15-, or 30-second counts are often used, since it is quite easy to convert to beats per minute (bpm) from these time segments by multiplying the pulse count by 10, 6, 4, or 2, respectively.
   c. Verify the heart rate obtained by palpation with that noted on the heart rate watch monitor during the same time period.
4. Be sure the subject is adequately recovered between testing procedures, if serial measures are to be completed on the same day. The heart rate should be back to pre-exercise resting levels before beginning the next submaximal test.

TREADMILL PROCEDURES

Single-stage and two-stage models have been used to estimate $\dot{V}O_{2max}$ from submaximal treadmill exercise (Mahar et al., 1985). In this laboratory exercise, a single-stage model based on the Bruce protocol (Bruce, Kusumi, and Hosmer, 1973) will be used. The same methodology can reasonably be applied to other treadmill protocols, as long as the energy costs of the exercise stages are known. For clarification, the terminology "single-stage" refers to the fact that $\dot{V}O_{2max}$ is predicted using only one submaximal data point, that being a heart rate value in the range of 115–155 bpm. As will be explained in the procedures below, it may be necessary for the subject to complete more than one stage of the Bruce protocol to reach the required heart rate. The specific procedures follow here.

1. With the subject standing near to observe, start the treadmill at a slow speed and demonstrate the safe and proper way to begin treadmill walking. (Review the procedures given in Laboratory 6 "Resting and Exercise Electrocardiography"). Thoroughly explain all procedures to the subject, then stop the treadmill.
2. Ask the subject to stand quietly with arms relaxed at the side on the unmoving belt of the treadmill. Measure the pre-exercise heart rate and blood pressure. Use palpation for the heart rate measure, and verify with the heart rate watch monitor. Record the values on the data sheet.
3. Ask the subject to grasp the support rails and straddle the treadmill belt. Carefully check to be sure the subject is completely off the belt, start it at a speed of 1.5 mph and 0% grade, and instruct the

subject to begin walking. Be sure to support the subject to help them maintain balance during the initial phase of treadmill walking.

4. As soon as the subject is walking comfortably, ask the subject to release the handrails and walk freely. Advance the treadmill to the first stage of the Bruce protocol (Table 8.3).
5. Measure the exercise heart rate the last 15 seconds of every minute of exercise and the blood pressure during the last minute of each stage; the measurement of heart rate takes precedence over blood pressure. Record the values of heart rate each minute and blood pressure each stage on the Treadmill Submaximal Test Worksheet found at the end of this laboratory chapter. It is very important that heart rate be measured and recorded accurately, since heart rate will be used to predict $\dot{V}O_{2max}$.
    a. The objective of the test is to reach a steady-state target heart rate between 115 and 155 bpm. This typically occurs in Stage 2 of this treadmill protocol, during the first 6 minutes of exercise.
    b. Finish the stage in which the target heart rate is achieved. Do not proceed to the next full stage if the subject's heart rate exceeds 135 bpm.
6. When the subject's heart rate measures between 115 and 155 bpm, continue the exercise protocol until the stage is completed, and then terminate the test at the end of the stage.
    a. Be sure to record the heart rate and blood pressure values corresponding to the last full stage of exercise on the Treadmill Submaximal Test Worksheet.
    b. Reduce the treadmill workload to 1.7 mph and 0% grade for a walking cool-down period.
    c. Measure blood pressure as soon as possible after exercise, and record this as the immediately post exercise (IPE) blood pressure on the worksheet.
    d. Continue to monitor the subject's heart rate and blood pressure each two minutes of recovery until the heart rate is less than 100 bpm.
    e. Although a walking recovery is recommended, if severe dizziness, fatigue, or other unusual complication arises, the subject should lie down. The subject should at no time be allowed to stand still until recovery is completed.
7. Calculate $\dot{V}O_{2max}$ on the Treadmill Submaximal Test Worksheet using the appropriate gender-specific formula shown below (Shephard, 1972). Note that $\dot{V}O_{2max}$ can be given either as ml · kg^{-1} · min^{-1} or as L · min^{-1} as long as the same units are used for SM$_{\dot{V}O2}$.

TABLE 8.3 — Stages of the Bruce protocol to be used for the single-stage model of submaximal exercise to estimate maximal oxygen uptake

Stage	Grade (%)	Speed (mph)	Time (min)	$\dot{V}O_2$* (ml · kg^{-1} · min^{-1})
I	10	1.7	1–3	14.4
II	12	2.5	4–6	20.2
III	14	3.4	7–9	30.2
IV	16	4.2	10–12	42.5

From Bruce, Kusumi, and Hosmer, 1973. $\dot{V}O_2$ for each stage is estimated from the last minute of a fully completed stage according to Foster et al., 1984.

Men

$$\dot{V}O_{2max} \ (ml \cdot kg^{-1} \cdot min^{-1}) = SM_{\dot{V}O2} \times [(HR_{MAX} - 61)/(HR_{SM} - 61)]$$

Women

$$\dot{V}O_{2max} \ (ml \cdot kg^{-1} \cdot min^{-1}) = SM_{\dot{V}O2} \times [(HR_{MAX} - 72)/(HR_{SM} - 72)]$$

Where: $SM_{\dot{V}O2}$ = submaximal $\dot{V}O2$ for the stage completed from Table 8.3.
HR_{MAX} = known maximal heart rate or predicted
HR_{SM} = submaximal heart rate achieved during the test

8. Refer to Table 8.1 for the subject's comparative fitness level.

STATIONARY CYCLE ERGOMETER

Stationary cycle ergometer single- and multiple-stage submaximal tests to predict $\dot{V}O_{2max}$ are more commonly used than treadmill tests. For example, the Åstrand-Rhyming test makes use of the heart rate response to a single-stage, 6-minute cycle ergometer submaximal workload to predict $\dot{V}O_{2max}$ (Åstrand and Ryhming, 1954). The multistage YMCA cycle ergometer protocol is also very popular and will be used in this laboratory (Golding, Myers, and Sinning, 1989). Three to four consecutive 3-minute stages are used in this cycle ergometer protocol to raise the heart rate to 110 to 150 bpm. Two heart rates in this range at two different workloads are required to predict $\dot{V}O_{2max}$. Also, in the procedures for this laboratory it is assumed that a friction-braked cycle ergometer, specifically a Monark® model, will be used. Other cycle ergometers can be used, as long as the workload or power output can be accurately set and measured. The specific procedures for this protocol follow here.

1. Preview the cycle ergometer protocol in Figure 8.4 and the Cycle Ergometer Submaximal Test Worksheet before beginning the laboratory. Note carefully in Figure 8.4 the decision point for advancing the workload after stage one depends on the subject's steady-state heart rate. Keep Figure 8.4 close at hand for easy reference during the testing procedure.

FIGURE 8.4 — Stationary cycle ergometer submaximal testing protocol. (Modified from "Y's Way to Physical Fitness," Golding, Myers, and Sinning, 1989).

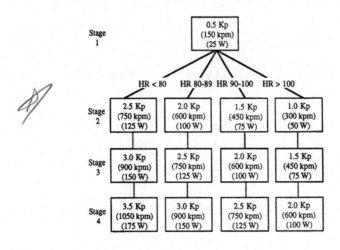

2. Demonstrate to the subject the proper riding technique and pedal frequency before the subject mounts the ergometer.
3. Ask the subject to be seated on the ergometer, hands on handlebars, and adjust the ergometer seat so that there is a slight bend (15°–25°) in the subject's leg when fully extended on the pedal.
4. Measure pre-exercise blood pressure and heart rate. Record these values on the Cycle Ergometer Submaximal Test Worksheet. See Figure 8.3 for palpating the pulse.
5. Start the submaximal exercise protocol.
    a. For warm-up, instruct the subject to pedal for 1 minute at 50 revolutions per minute (rpm) and at a very light resistance of 0.25 kp (12.5 W).
        i. A continuous pedal frequency of 50 revolutions per minute (rpm) in every stage is critically important for this procedure. Use the rpm readout on the ergometer if so equipped and verified as accurate, or set a metronome to emit a sound/light signal 100 times per minute. Instruct the subject to pedal first with one leg, then the other in time with the metronome signal. This will ensure one complete pedal revolution with each two metronome signals giving a pedal frequency of 50 rpm.
    b. Precisely after 1 minute of warm-up, increase the resistance to 0.5 kp for stage 1, and coach the subject to maintain 50 rpm pedal frequency. The initial work rate will be relatively light at 150 kpm · min^{-1} or 25 W.
        i. The duration of this initial exercise stage and all succeeding stages must be exactly 3 minutes.
        ii. Measure blood pressure at the 1.5-minute mark.
        iii. Measure heart rate twice during this stage, once each the last 30 seconds of minutes 2 and 3. If these two heart rates differ by more than 5 bpm, the test stage should be extended for an additional minute, or until a stable, steady-state heart rate value is obtained.
        iv. Be sure to accurately record the heart rate and blood pressure information on the Cycle Ergometer Submaximal Test Worksheet.
    c. **DECISION POINT:** Refer to Figure 8.4. Use the subject's heart rate at the end of stage 1 to choose the workload column to follow for all subsequent workloads. For example, if the subject's stage 1 heart rate ranged from 90 to 100 bpm, the technician would set the second workload at 1.5 kp (450 kpm· min^{-1} or 75 W), and successive workloads at 2.0 kp and 2.5 kp for stages 3 and 4, respectively.
    d. Repeat blood pressure and heart rate measurements during each exercise stage as described above for stage 1, and carefully record the values on the Cycle Ergometer Submaximal Test Worksheet.
    e. Continue the protocol shown in Figure 8.4 *until*:
        i. Heart rates recorded in *two successive* stages measure between 110 and 150 bpm. For many individuals, this will occur in stages 2 and 3; in this case, stage 4 would not be necessary.
            (1) Complete the stage in which the heart rate objective is reached.
            (2) Do not advance to the next stage if the heart rate exceeds 150 bpm.
        ii. The subject requests to stop or can no longer maintain the 50-rpm pedal frequency.
6. Once the test is completed, initiate the cool-down period. Reduce the resistance to 0.5 kp and allow the subject to pedal an additional 1 to 2 minutes, or until the heart rate falls below 100 bpm and blood pressure normalizes.

7. Estimate $\dot{V}O_{2max}$ using the prescribed four-step process (Golding, Myers, and Sinning, 1989) by following the instructions given on the Cycle Ergometer Submaximal Test Worksheet, and using the Cycle Ergometer Data Graph. Alternately, a graphing or charting computer program may be used to construct this graph.

STEP BOX OR BENCH STEP TEST

The following procedure to predict $\dot{V}O_{2max}$ from step box or bench stepping was developed by McArdle and colleagues (1972) for college-age men and women. The standard error of the prediction was reportedly about ± 16%.

1. Unlike the treadmill and cycle ergometer tests, this procedure does not provide for a gradual increase in intensity. Therefore, before starting this procedure, guide the subject through some gentle stretching exercises and brief walking in place for warm-up.
2. The step-box height used in this protocol is 16.25 in (41.25 cm), and the step rate is 24 steps-per-minute (spm) for men and 22 spm for women. (See Figure 8.5.)
    a. To control the step rate, set the metronome to emit a sound/light signal 96 times-per-minute for men and 88 for women.
3. The technician must carefully locate the carotid or radial pulse, marking it with a felt-tip pen, before starting the test so that it can be quickly located during recovery. It is *critical* that the pulse be located within 5 seconds of stopping the step exercise so that accurate early recovery pulse counts can be obtained.
    a. Note: If a heart rate watch monitor is used, it *must* show a visual or audible signal for each single heartbeat, which can be counted by the technician during recovery. Reading the heart rate off the watch during recovery is not acceptable for this procedure, since the watch reading will reflect an average of several heart beats, not necessary over the 15 seconds required for this $\dot{V}O_{2max}$ prediction.
4. The technician should thoroughly describe the procedure to the subject, demonstrate the proper step procedure, and detail how and when the heart-rate measures will be taken. The step-box procedure to demonstrate and explain is as follows.

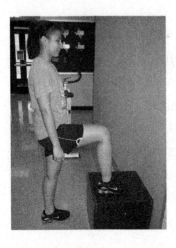

FIGURE 8.5 — Starting position for step-box submaximal test.

   a. Start the metronome at the proper cadence, and place it near the step box to be easily heard and seen during the stepping exercise.
   b. Stand comfortably facing the step box or bench, feet shoulder width apart, and toes of shoes close to the bottom of the step box.
   c. Listen to or watch the metronome signal until the rhythm is acquired. Begin by stepping up onto the top of the step box with one foot in time with the metronome signal, extending the knee straight in the step-up motion. Follow immediately with a step up with the opposite foot on the second signal, again extending the knee so that both feet are now on top of the step box, both knees fully extended straight, and the body posture standing fully upright. On the third metronome signal, step down with the first foot, then down with the opposite foot on the fourth signal. It is often helpful to verbalize "Up, Up, Down, Down," in rhythm with the metronome signal as the step sequence is executed to aid with acquiring the step rhythm.
   d. Reverse legs on every-other stepping series to avoid excessive fatigue in one leg.

5. After demonstrating the procedure, ask the subject to assume the standing position facing the step box. Measure pre-exercise heart rate and blood pressure, then record these values on the Step Test Worksheet.
6. Start the metronome at the proper cadence. Alert the subject that on the command to "Begin Exercise," he or she is to start the stepping exercise as was demonstrated.
   a. With the begin exercise command, start a stopwatch to time the length of the exercise.
7. The subject is to continue exercise in cadence with the metronome and without interruption for 3 minutes. No rest periods are allowed. If the subject stops exercise for any reason before the 3 minutes are completed, the test data cannot be used to predict $\dot{V}O_{2max}$.
8. At the end of 3 minutes, instruct the subject to stand still and upright with both feet on the floor next to the step box or bench.
   a. Immediately reset the stopwatch to 0, and restart it.
   b. Marking the time with the stopwatch, *quickly* locate the pulse within the first 5 seconds after stopping exercise.
   c. Count the number of pulse beats for 15 seconds beginning with the fifth second after stopping exercise and continuing to the twentieth second. Convert this 15-second count to beats per minute by multiplying by 4. Record this value on the Step Test Worksheet.
   d. After the pulse count has been taken, measure and record a recovery blood pressure.
   e. Allow the subject to cool down by walking slowly around the room or in place.
9. Use the data to predict $\dot{V}O_{2max}$ by completing the calculations portion of the Step Test Worksheet. Note that the units for the predicted $\dot{V}O_{2max}$ value will be ml· kg^{-1} · min^{-1}. To convert to L· min^{-1}, simply multiply by the body weight in kilograms.

QUESTIONS AND ACTIVITIES

1. List and explain potential sources of error in predicting $\dot{V}O_{2max}$ using the three methods described in this laboratory exercise. How could these sources of error be controlled?
2. Which submaximal protocol would you choose to apply in the following situations. Provide reasons to support your choices.
   a. College-age men and women enrolled in a fitness and conditioning class.
   b. Elderly men attending a senior-citizens' fitness program.
   c. Men and women enrolled in a step-box aerobic fitness class.
3. Would you expect $\dot{V}O_{2max}$ to be overestimated or underestimated in well-trained cyclists if you used the YMCA cycle ergometer protocol to test them? Give reasons to support your answer.
4. In all submaximal protocols for predicting $\dot{V}O_{2max}$, it was emphasized that heart rate should be kept in a range between 110 and 150 bpm. Why is this important?
5. Give reasons for preliminary screening of subjects before they are exercise tested.
6. Did blood pressure respond to exercise as you would expect? Why or why not?

QUESTIONS AND ACTIVITIES FOR GOING FURTHER

1. Using data combined from all class members, apply statistical procedures to:
   a. Determine the strength of the relationships between the methods of predicting $\dot{V}O_{2max}$.
   b. Test for significant differences in $\dot{V}O_{2max}$ for the class members predicted by these three submaximal test methods.
   c. Provide reasons for any agreement or disagreement in the results of the three methods.

2. Choose any member of your lab group to be your client, and suppose he or she wants to lose 15 pounds in the next 6 months. The subject is currently weight stable, with a caloric intake that balances expenditure. Use the data from the subject's cycle ergometer test to determine the number of minutes each week he or she would need to exercise at 70% of $\dot{V}O_{2max}$ to lose this amount of body weight. Assume your client wishes to lose the same amount of weight each week, and he or she does not increase caloric intake to cover the energy expended in exercise.

REFERENCES

Åstrand, P. O. and I. A. Ryhming. "A Nomogram for Calculation of Aerobic Capacity (Physical Fitness) from Pulse Rate during Submaximal Work." *J. Appl. Physiol.* 7 (1954): 2, 218–21.

Åstrand, P. and P. Åstrand. *Textbook of Work Physiology: Physiological Bases of Exercise.* Champaign, IL: Human Kinetics, 2003.

Brouha, L. "The Step Test: A Simple Method of Measuring Physical Fitness for Muscular Work in Young Men." *Res. Q.* 14 (1943): 1, 31–36.

Bruce, R. A., F. Kusumi, and D. Hosmer. "Maximal Oxygen Intake and Nomographic Assessment of Functional Aerobic Impairment in Cardiovascular Disease." *Am. Heart J.* 85 (1973): 4, 546–62.

Cooper, K. H. "A Means of Assessing Maximal Oxygen Intake: Correlation between Field and Treadmill Testing." *JAMA* 203 (1968): 3, 201–04.

Ekblom, B., P. O. Åstrand, B. Saltin, J. Stenberg, and B. Wallstro. "Effect of Training on Circulatory Response to Exercise." *J. Appl. Physiol.* 24 (1968): 4, 518–28.

Foster, C, A. S. Jackson, M. L. Pollock, M. M. Taylor, J. Hare, S. M. Sennett, J. L. Rod, M. Sarwar, and D. H. Schmidt. Generalized Equations for Predicting Functional: Capacity from Treadmill Performance." *Am. Heart J.* 107 (1984): 6, 1229–34.

Gellish, R. L., B. R. Goslin, R. E. Olson, A. McDonald, G. D. Russi, and V. K. Moudgil. "Longitudinal Modeling of the Relationship Between Age and Maximal Heart Rate." *Med. Sci. Sports Exerc.* 39 (2007): 5, 822–29.

Golding, L. A., C. R. Myers, and W. E. Sinning. *Y's Way to Physical Fitness: The Complete Guide to Fitness Testing and Instruction.* Champaign, IL: Published for YMCA of the USA by Human Kinetics Publishers, 1989.

Heyward, V. H. *Advanced Fitness Assessment and Exercise Prescription.* Champaign, IL: Human Kinetics, 1998.

Mahar, M. T., A. S. Jackson, R. M. Ross, J. M. Pivarnik, and M. L. Pollock. "Predictive Accuracy of Single and Double Stage Sub Max Treadmill Work for Estimating Aerobic Capacity." *Med. Sci. Sports Exerc.* 17 (1985): 2, 206–7.

McArdle, W. D., F. I. Katch, G. S. Pechar, L. Jacobson, and S. Ruck. "Reliability and Interrelationships between Maximal Oxygen Intake, Physical Work Capacity and Step: Test Scores in College Women." *Med. Sci. Sports Exerc.* 4 (1972): 4, 182–86.

Myers, J, R. Arena, B. Franklin, I. Pina, W. E. Kraus, K. McInnis, G. J. Balady, Council Clin Cardiology and Council Cardiovasc Nursing. "Recommendations for Clinical Exercise Laboratories: A Scientific Statement From the American Heart Association." *Circulation* 119 (2009): 24, 3144–61.

Rockport Walking Institute. Rockport Fitness Walking Test. 1986.

Shephard, R. J. *Alive Man! The Physiology of Physical Activity.* Springfield, IL.: Thomas, 1972.

Thompson, W. R, N F. Gordon, and L. S. Pescatello (eds.). *ACSM's Guidelines for Exercise Testing and Prescription.* Philadelphia: Lippincott Williams & Wilkins, 2010.

Informed Consent For Submaximal Exercise Testing

(Modified from Thompson, Gordon, and Pescatello's "ACSM's Guidelines for Exercise Testing and Prescription," 2010.)

1. **Explanation of Test Procedure**

    You will perform a <u>submaximal</u> exercise test on a stationary cycle ergometer, a motor-driven treadmill, or a step box. The exercise intensity will begin at a level you can easily accomplish and will be advanced no more than two stages of difficulty, depending on your exercise ability, and will be stopped at a moderate level of difficulty. The testing personnel may stop the test at any time because of signs of fatigue or symptoms that may not be considered normal. And you may choose to stop exercise whenever you wish because of fatigue or discomfort, or if you simply do not wish to continue exercise.

2. **Possible Risks and Discomforts**

    There is the possibility of certain physical changes occurring during the test, some of which may be harmful. These potential changes include abnormal blood pressure, fainting, and disorders of heart beat. It is very rare, but there is also a possibility of heart attack, stroke, or death during exercise. Every effort will be made to minimize the occurrence of these problems by evaluating preliminary information you have provided related to your health and physical fitness, and by careful observations during testing. Emergency equipment is available, and para-medical emergency service is within _____ miles of this testing facility. Also, trained personnel are available to deal with unusual situations which may arise.

3. **Your Responsibilities**

    To the best of your knowledge you are responsible to truthfully report information about your medical history, current health status, and previous experiences of symptoms which may be related to your heart. These include symptoms such as shortness of breath with light physical activity, pain, pressure, tightness, or heaviness in the chest, neck, jaw, back, and/or arms with physical effort. These conditions may affect the safety of your exercise test. It is very important that you promptly report any of these and any other unusual feelings at any time during the exercise test. You should also report to the testing staff all medications prescribed by your doctor and those you take without prescription, the time you last took your medicines, and whether you took them today.

4. **Benefits You Can Expect**

    The results of the test will be used to provide you with information about your current level of physical working capacity, that is, your physical fitness and cardiorespiratory endurance capacity. The information will also aid in defining an amount of physical activity that is safe and that will help improve your physical fitness. This type of test is not designed to diagnose an illness, or evaluate the effect of any medical treatment you may have had.

5. **Inquiries about the Test**

    You are encouraged to ask any questions you have about this submaximal exercise test before you begin, and after the test is completed. You should not begin this test until you have your questions answered to your satisfaction.

6. **Use of Medical Records**

    The information obtained during this submaximal test will be treated as privileged and confidential in compliance with the Health Insurance Portability and Accountability Act of 1996. No information will be released or revealed to anyone other than your referring physician without written consent from you. You should understand, however, that the information obtained may be used for statistical analysis or scientific purposes with your right of privacy retained.

7. **Consent for the Test**

   I voluntarily consent and give my permission to perform this submaximal exercise test designed to help me understand more about my physical fitness and health. I understand that I am free to deny consent and to stop the test at any point I so desire.

I have read this consent form, I understand the test procedures that I will perform, and I have been provided an explanation of the possible risks and discomforts. Knowing these risks and discomforts, and having had an opportunity to ask questions that have been answered to my satisfaction, I consent to participate in this submaximal exercise test.

_____   _____

Signature of Participant   Date

_____   _____

Signature of Witness   Date

SUBMAXIMAL TESTING READINESS QUESTIONNAIRE

Directions: This questionnaire is designed to obtain information that will help determine whether or not submaximal exercise testing will be safe for you. Please complete the Personal Information and Health Information sections of the form. (Modified from Thompson, Gordon, and Pescatello, *ACSM's Guidelines for Exercise Testing and Prescription,* Philadelphia: Lippincott Williams & Wilkins, 2010.)

Personal Information

Name:_____ Age:_____ Gender: M_____ F_____

Address:_____ Date:_____/_____/_____

Phone:_____(work)   _____(home)   _____(cell)

Health Information

Please check <u>ONLY</u> those which apply to you.

___ 1. My doctor has told me that I have heart problems, and I should only do the physical activities recommended by my doctor.

___ 2. I often feel pains in my chest, or have abnormal heart beats when I work or exercise.

___ 3. In the past month I have had chest pains when I was not doing any physical work or exercise.

___ 4. I have times when I feel faint, dizzy, or even passed out, or I have lost my balance when I became dizzy.

___ 5. My doctor has told me that I have high blood pressure or high cholesterol.

___ 6. I have bone or joint problems (e.g., back, knee, or ankle) that keep me from exercising or might be made worse if I exercise.

___ 7. I <u>do not</u> regularly do physical activity hard enough to make me sweat.

___ 8. I regularly take medications prescribed by my doctor.

   a. Give purpose: _____

   b. Name/Dose: _____

___ 9. I smoke or use tobacco products regularly.

___ 10. Women: I am now or I believe I might be pregnant.

___ 11. There are other reasons why I should not exercise.

   a. List reasons: _____

NOTE: If any items are checked, the exercise test should be postponed until further review by a qualified professional. Medical clearance may be necessary before the individual begins to exercise or takes part in an exercise test.

TREADMILL SUBMAXIMAL TEST WORKSHEET

Name: _____ Age: 20 Date: _____

Weight: 105 lb _____ kg Actual or Age-Predicted Max HR: 200 bpm

Treadmill Test Data

Resting Pre-Exercise	HR (bpm)	_____	Comments:			
	SBP/DBP (mmHg)	___/___				

Exercise	Stage	Speed (mph)	Grade (%)	HR (bpm) 1 min	2 min	3 min	SBP/DBP (mmHg)
	1	1.7	10	100	104	100 bpm	
	2	2.5	12	116	116	112	
	3	3.4	14	120	128	128	
	4	4.2	16		120	128	

Recovery	Min	HR	BP	Comments:
	IPE			
	2	76		
	4			

Calculations to Predict $\dot{V}O_{2max}$ from Treadmill Data

$SM_{\dot{V}O_2}$ = submax $\dot{V}O_2$ from Table 3 HR_{MAX} = max heart rate HR_{SM} = submax heart rate end test

Men

$\dot{V}O_{2max} = SM_{\dot{V}O_2}$ _____ $\times$ [(HR_{MAX} _____ $- 61$) $\div$ (SM_{HR} _____ $- 61$)] = _____

Women

$\dot{V}O_{2max} = SM_{\dot{V}O_2}$ _____ $\times$ [(HR_{MAX} _____ $- 72$) $\div$ (SM_{HR} _____ $- 72$)] = _____

CYCLE ERGOMETER SUBMAXIMAL TEST WORKSHEET

Subject name: Jason Forrow		Age (yr): 25	Wt (kg): 97	Wt (lb): 215
Pred Max HR: 195	Rest HR:	Rest BP:	Meds/Limitations:	

Stage	Resistance (kp)	HR (bpm)	BP (mm/Hg)	Comments:
1	0.5 kp	88		
2	2.0 kp	116		
3	2.5 kp	120		
4				

Instructions for Cycle Ergometer Data Graph to Predict $\dot{V}O_{2max}$

STEP 1
Draw a horizontal line across the graph at the maximum HR. If unknown, estimate by 220-age.

STEP 2
Plot the HR measured the third minute of the final two stages against their corresponding workloads.

STEP 3
Draw a line through both points and extend to the point of intersection with the max HR line.

STEP 4
Drop a vertical line from intersection to the x-axis and read the predicted max workload and $\dot{V}O_{2max}$.

Predicted Max Workload (kpm·min^{-1}) _____ (kpm·min^{-1})   Predicted $\dot{V}O_{2max}$ (L·min^{-1}) _____ (L·min^{-1})

Cycle Ergometer Submaximal Data Graph

Workload (kpm·min^{-1})	150	300	450	600	750	900	1050	1200	1350	1500	1650	1800	1950	2100
$\dot{V}O_{2max}$ (L·min^{-1})	0.6	0.9	1.2	1.5	1.8	2.1	2.4	2.8	3.2	3.5	3.8	4.2	4.6	5.0
Kcal (Kcal·min^{-1})	3.0	4.5	6.0	7.5	9.0	10.5	12.0	14.0	16.0	17.5	19.0	21.0	23.0	25.0

0.5 × 50 rpm × 6 m/rev

kpm = kp × RPM × m/rev

STEP TEST WORKSHEET

Subject name:			Age (yr): 20	Wt (kg): 70.9	Wt (lb):
Pred Max HR: 200	Rest HR: 68	Rest BP:	Meds/Limitations:		

EXERCISE: Perform stepping exercise on a step-box or bench 16.25" high for 3 minutes at a step rate of 24 steps per minute for men and 22 for women

Comments:

Recovery Measures

Pulse Count in 15 sec (PC_{15s}) = __31__

HR_{REC} (bpm) = $PC_{15s} \times 4$ = __124__ (bpm)

BP (mmHg) = ____/____

Calculations to Predict $\dot{V}O_{2max}$ from Step Test Data

HR_{REC} = recovery heart rate expressed as bpm calculated above

Men

$\dot{V}O_{2max} = 111.33 - (0.42 \times HR_{REC}$ ____ $) =$ ____ $ml \cdot kg^{-1} \cdot min^{-1}$

Women

$\dot{V}O_{2max} = 65.81 - (0.1847 \times HR_{REC}$ ____ $) =$ ____ $ml \cdot kg^{-1} \cdot min^{-1}$

CHAPTER 9

THE SYMPTOM-LIMITED MAXIMAL GRADED EXERCISE TEST FOR CLINICAL AND SPORTS MEDICINE APPLICATIONS

OUTCOME OBJECTIVES

Following successful completion of this exercise, the student should be able to:

1. List and describe important preliminary screening procedures and identify contraindications for the symptom-limited maximal graded exercise test (SLM-GXT).
2. Measure the ECG and blood pressure at rest, during submaximal, and during maximal exercise, and recognize abnormal symptoms.
3. Safely conduct a symptom-limited maximal graded exercise test on a motor driven treadmill or stationary cycle ergometer.
4. Recognize reasons for stopping an SLM-GXT.
5. Evaluate as normal or abnormal the exercise ECG and physiologic results of the maximal SLM-GXT, including blood pressure, heart rate, $\dot{V}O_2$, and maximal working capacity ($\dot{V}O_{2max}$).

INTRODUCTION

The measurement of the physiologic responses to exercise has long been used to obtain information related to an individual's physical performance limits, general physical health, or response to medical therapy. Over the years a number of procedures have been employed to produce the exercise stimulus, from simple stepping and walking protocols to the use of automated ergometers and motorized treadmills. The most common procedure employed today in clinical medicine and human performance assessment laboratories is broadly termed the graded exercise test (GXT) or, alternatively, the exercise stress test. By definition, the GXT is a procedure whereby physical work is incrementally increased in a controlled manner while various physiologic responses are measured. Most frequently, the motorized treadmill or stationary cycle ergometer are the instruments of choice to provide the exercise stress. The GXT protocol may call for a less than maximal work effort, in which case it is termed a submaximal GXT. However, more information can be obtained if the protocol requires that the subject exert a maximal effort until limited by fatigue or some other symptom; such a test is often termed a symptom-limited maximal GXT (SLM-GXT). The procedures for a SLM-GXT using a motor-driven treadmill will be described in this laboratory exercise. The general procedures are essentially the same for a cycle ergometer, except that the work increments may be different, depending on the protocol. The GXT has proven to be relatively safe procedure with risks of a serious cardiac event resulting in death or requiring hospitalization reportedly about 1 per 10,000 when established techniques, trained staff, and appropriate monitoring are used. The estimates of sudden cardiac death during a SLM-GXT range from 0 to 5 per 100,000 tests (Meyers, Arena, Franklin, Pina, Kraus, and McInnis, 2009).

Although in practice the actual procedures may be very similar, two types of GXTs have been defined based on the primary use of the information obtained from the test (Thompson, Gordon, and Pescatello, 2010). In clinical settings a *Diagnostic* GXT, which includes the exercise 12-lead electrocardiogram (ECG), is used as a diagnostic or prognostic tool in the evaluation of suspected or established cardiovascular disease. By contrast, a *Functional* GXT is employed in other situations, such as in physical fitness and rehabilitation programs of hospitals, corporations, universities, and sports medicine centers, to measure functional capacity, also variously termed cardiorespiratory endurance, maximal working capacity, or aerobic capacity. The information obtained from a Functional GXT is then used as the basis for an exercise prescription to improve health, rehabilitate those with cardiac and pulmonary disease, or maximize sport performance. It should be evident that no matter what the primary purpose of the SLM-GXT, the data obtained can provide both diagnostic and functional information. For example, latent cardiovascular disease may be uncovered in an apparently healthy individual during a Functional SLM-GXT.

Whatever the primary application of the SLM-GXT, the procedures generally call for the measurement of several physiologic variables. Most often a 12-lead ECG is recorded and blood pressure is measured at intervals coinciding with changes in the workload, termed stages in the GXT protocol. In pulmonary-limited individuals, blood oxygen saturation is often measured using an ear or finger oximeter. These variables are closely observed during exercise for abnormal changes which may signal that the GXT should be stopped (see Table 9.1).

In both diagnostic and functional applications of the SLM-GXT, assessment of $\dot{V}O_{2max}$ is generally of primary importance. Physiologically, $\dot{V}O_{2max}$ is the single most important measure of physical working capacity, and is usually defined as the maximal amount of oxygen the body can take up and use during maximal exertion. It is generally reported in one of two ways, either as liters of oxygen used per minute ($L \cdot min^{-1}$) or relative to body weight as milliliters of oxygen used per kilogram of body weight per minute ($ml \cdot kg^{-1} \cdot min^{-1}$). When comparing physical fitness levels between individuals, the relative measure is used. Also, it is often useful to convert relative oxygen uptake to a unitless value termed a metabolic equivalent (MET). A MET is defined as the average oxygen uptake measured in men and women at quiet rest, determined experimentally to be $3.5 \; ml \cdot kg^{-1} \cdot min^{-1}$. Conversion to METs is done by dividing $\dot{V}O_2$ expressed as $ml \cdot kg^{-1} \cdot min^{-1}$ by $3.5 \; ml \cdot kg^{-1} \cdot min^{-1}$. METs are frequently used in clinical settings to express an individual's maximal working capacity and to rate the intensities of certain types of physical tasks, therefore useful in prescribing exercise. For example, it is unlikely that it would be recommended for a cardiac patient with a maximal working capacity of 7 METs to play tennis, which is a 6.5 MET activity, yet an exercise prescription may include cart golf, since this is a 2–3 MET activity.

Direct measurement of $\dot{V}O_{2max}$ can easily be accomplished, but it requires expensive and sophisticated equipment, as well as trained technicians. Rapid response automated metabolic gas analyzers are widely available to measure $\dot{V}O_2$ breath-by-breath, and display the oxygen uptake throughout each stage of the GXT. $\dot{V}O_{2max}$ should be measured directly when data are to be used for research purposes, or when very precise measurements of oxygen uptake are required, such as in training elite athletes. For direct measurement procedures, the student is referred to the Laboratory 5 entitled *Measurement of Oxygen Uptake and Energy Expenditure*. However, for general applications and clinical assessment of fitness, such precision is often not necessary. $\dot{V}O_{2max}$ is often estimated in clinical practice rather

TABLE 9.1 – Reasons for stopping a SLM-GXT in low-risk adults and in clinical testing

1. Subject requests to stop.
2. Failure of the equipment or technical difficulties, including uninterpretable ECG due to artifact or electrode detachment.
3. Beginning of chest pain interpretable as angina or anginalike symptoms, or increasing chest pain, angina >3 on 4-point angina scale.
4. Subject exhibits symptoms of severe fatigue, nervous system, or circulatory problems such as nausea, ataxia, lightheadedness, near syncope, confusion, cyanosis, pallor, leg cramps, claudication.

5. Abnormal blood pressure responses to exercise including:
  a. Significant drop ≥10 mmHg in systolic blood pressure with increase in workload, especially if accompanied by other signs of ischemia.
  b. Failure of systolic blood pressure to rise appropriately with increase in workload.
  c. Excessive blood pressure rise: systolic >250 and diastolic >115 mmHg.

6. Respiratory symptoms such as severe shortness of breath, wheezing.

7. Abnormal ECG changes during exercise including:
  a. Sustained ventricular tachycardia is absolute reason to stop.
  b. ST-segment elevation ≥1mm in leads other than V_1 or aVR.
  c. Abnormal QRS axis shift.
  d. Flat or downsloping ST-segment depression >2 mm.
  e. Onset of 2nd or 3rd degree heart block.
  f. Multiform PVCs, intermittent ventricular tachycardia, R-on-T PVCs, or increasing ventricular ectopy.
  g. Superventricular tachycardia or bradyarrhythmias.
  h. Exercise-induced bundle branch blocks or intraventricular conduction delays indistinguishable from ventricular tachycardia.

8. Failure of the heart rate to increase appropriately with an increase in workload in the absence of heart rate limiting drugs.

Note: in clinical settings with physician supervision, the SLM-GXT may be continued beyond these reasons to stop if the benefits are judged by the physician to outweigh the risks.
Modified from Thompson, Gordon, and Pescatello (eds), 2010.

than measured directly, and a number of protocol-specific prediction equations have been developed to accomplish this task with some degree of accuracy (Table 9.2). In this laboratory, specific equations in Table 9.2 will be used to calculate $\dot{V}O_{2max}$ from the total time the subject is able to walk/jog on the treadmill. The student is encouraged to apply direct measurement procedures, if equipment is available, for an interesting comparison of direct and indirect measures of $\dot{V}O_2$ during exercise.

Knowledge of bioenergetics and circulatory and respiratory physiology is foundational for understanding the significance of $\dot{V}O_{2max}$ and its relationship to physical working capacity. Energy for muscular contraction during exercise is produced in three general ways:

1. The splitting of the high energy phosphates, adenosine triphosphate (ATP) and creatine phosphate (CP), which are stored at the contractile site and immediately available to regenerate the ATP consumed during short bouts of high-intensity work (about 15 seconds).

TABLE 9.2 – Protocol-specific equations for estimating $\dot{V}O_{2max}$ (ml · kg^{-1} · min^{-1}) using total exercise time in minutes (Treadmill) or maximal workload (Cycle ergometer)

PROTOCOL	REFERENCE	EQUATION
Bruce	Foster, 1984	Generalized Equation for Men and Women $\dot{V}O_2 = 14.8 - 1.379(TIME) + 0.451(TIME^2) - 0.012(TIME^3)$
Balke	Pollock, 1976	Active and Sedentary Men $\dot{V}O_2 = 1.44(TIME) + 14.99$
Balke*	Pollock, 1982	Active and Sedentary Women $\dot{V}O_2 = 1.38(TIME) + 5.22$
Naughton	Foster, 1983	Generalized Equation for Men and Women $\dot{V}O_2 = 1.61(TIME) + 3.60$
Leg Cycle Ergometer	Thompson, 2010	Generalized Equation for Men and Women $\dot{V}O_2 = $ (Workrate in kpm · min^{-1} × 1.8 ml O$_2$ · kpm^{-1} ÷ kg body mass) + 7.0 (ml O$_2$ · kg^{-1} · min^{-1})

NOTE: The unit for TIME is minutes and the decimal equivalent of seconds; e.g., 6 min and 30 seconds would be 6.5 minutes.
* Balke protocol modified for women: Speed constant 3.0 mph; 3 min stages starting at 0% grade and increasing 2.5% every stage thereafter.

2. Glycolysis, which may produce ATP anaerobically by reducing pyruvate to lactate, fuels moderate to high intensity activity lasting up to 3 minutes.
3. Aerobic metabolism, by which ATP is regenerated from the oxidation of pyruvate and fatty acids to fuel activity for extended periods of time.

All three processes occur concurrently, and the energy derived from each will vary according to the intensity of the exercise. However, the ability to perform work for extended periods of time (more than about 3 minutes) is dependent almost exclusively on aerobic metabolism, and this energy-producing process is limited by the maximal rate at which oxygen can be taken up and utilized, that is, by $\dot{V}O_{2max}$.

Physiologically, $\dot{V}O_{2max}$ is dependent upon two general factors. First, oxygen *supply* to muscle tissue may prove limiting. If the combined ability of the respiratory and cardiovascular systems to take up atmospheric oxygen and transport it to the active muscle tissues is impaired by disease or diminished through detraining, then $\dot{V}O_{2max}$ will be reduced. With appropriate physical training, the efficiency and capacity of these systems can be improved, and $\dot{V}O_{2max}$ will likewise increase. Second, oxygen *utilization* by the active tissue affects $\dot{V}O_{2max}$, and, consequently, the amount of work that can be accomplished aerobically. This factor is related to the metabolic machinery of the muscle tissue itself, which can adapt to training by increasing its capacity to produce energy aerobically. The SLM-GXT, then, as a test of $\dot{V}O_{2max}$, provides information that can be used to evaluate the elements of physical fitness shown in Table 9.3. Note that obesity is evaluated indirectly by the GXT. Fat tissue is essentially metabolically inert with respect to $\dot{V}O_{2max}$ measurements. Yet when $\dot{V}O_{2max}$ is expressed relative to body weight, the calculated value will be reduced proportionally to excess

TABLE 9.3 – Elements of physical fitness which can be assessed using the SLM-GXT

1. Cardiovascular function
2. Respiratory function
3. Oxidative capacity of muscle
4. O_2 carrying capacity of blood
5. Muscular strength and endurance
6. Obesity

fat weight. Since the relative $\dot{V}O_{2max}$ value is most often used as the standard for comparison when evaluating physical fitness, an obese individual will appear to be less fit compared to his or her non-obese counterparts.

Once the SLM-GXT is completed, the data should be organized in a final summary report for ease of interpretation. Answers to the questions posed in Table 9.4 will be provided by the GXT data, and should be included in some fashion in a summary report, although the relative importance of the items may vary depending on either a diagnostic or functional objective for the SLM-GXT.

Equipment Needed

12-lead ECG machine with paper recorder (calibrated 1 mV = 10 mm, paper speed 25 mm · sec^{-1})

Motor-driven treadmill (alternately stationary cycle ergometer)

Electric hair trimmer or disposable dry razor

Silver-silver chloride electrodes (disposable)

ECG prep abrading tape or gauze and alcohol

Sphygmomanometer, blood pressure cuff, and stethoscope

Metabolic cart optional for direct measurement of $\dot{V}O_{2max}$

TABLE 9.4 – General questions to be answered from the SLM-GXT data

1. Were normal test end-points achieved indicating a true maximal effort?
   a. At least 85% of age-predicted maximum heart rate achieved?
   b. A Borg rating of perceived exertion of at least 17 (very hard) reported?
   c. Limiting symptoms reported to be general fatigue, specific muscle fatigue, or hyperpnea.

2. Were ECG responses normal?
3. Were abnormal symptoms noted, such as angina, severe dyspnea, or claudication?
4. Did the blood pressure, heart rate, and rate-pressure product respond to exercise as expected?
5. Was $\dot{V}O_{2max}$, whether calculated or measured, above or below predicted?
6. Based on the GXT data, would it be safe for the individual to begin or continue an exercise program?   a. At what level of intensity?

TECHNIQUES AND PROCEDURES

The procedures in this laboratory comply in general with those recommended by the American Heart Association (Gibbons et al., 1997; Myers, et al., 2009) and by the American College of Sports Medicine (Thompson, Gordon, and Pescatello, 2010). In the SLM-GXT procedures which follow, broad references will be made to the skills acquired by the student in completing Laboratories 6 and 7. Direct measures of $\dot{V}O_2$ with a computerized metabolic cart (see Laboratory 5 "Measurement of Oxygen Uptake and Energy Expenditure") are not commonly performed in a clinical setting, but often are included as part of SLM-GXT protocols in human performance laboratories where more precise measures of cardiovascular capacity are required for research or sport medicine applications. The logical time-points in the SLM-GXT procedures to make these measures will be highlighted without providing specific procedural details for those who have this laboratory capability and wish to add these valuable measures to this laboratory experience.

Although it is possible for all the technical procedures to be performed by one person, it is more often the case that an SLM-GXT is conducted by a team of two or more individuals. The data collected during a GXT are extensive, and the recording forms require multiple entries before, during, and after exercise, more efficiently completed by more than one individual. In a clinical setting, the team usually consists of a physician and clinical exercise physiologist or nurse. When the test does not warrant the direct supervision of a physician, two clinical exercise physiologists, nurses, or other trained and certified personnel may conduct the test (Myers et al., 2009; Thompson, Gordon, and Pescatello, 2010). The procedures outlined below are designed to be completed by a team of three persons. Note that if $\dot{V}O_2$ measures will be made with a computerized metabolic cart, modifications in these personnel task assignments will be necessary and may require an additional team member. Computerized metabolic carts may also include the option of recording the ECG, which reduces the staff requirement to obtain these multiple measures during exercise. One method of dividing the SLM-GXT tasks among testing personnel follows here.

1. **ECG Technician**—preps and monitors the ECG, operates the controls for the treadmill (or cycle ergometer), records data, makes necessary data reduction calculations, ensures accuracy of the SLM-GXT data, and compiles a GXT final report.
2. **Blood Pressure Technician**—measures resting and exercise blood pressures, demonstrates procedures to the subject, visually observes the subject for any symptoms of problems and for signs of fatigue near maximal exertion, maintains verbal communication with the subject, and acquires Rating of Perceived Exertion (RPE) scores throughout the GXT, and assists in data reduction and the final report.
3. **Subject**—serves as the subject to complete the SLM-GXT.

GENERAL PROCEDURES TO PREPARE FOR THE SLM-GXT

1. Communicate to the subject in advance of arrival:
    a. To avoid heavy meals at least 4 hours before arriving at the testing facility, and report well-hydrated.
    b. Information about the nature of the exercise test, what type of exercise will be performed, and what measures will be taken.
    c. Instructions to dress in clothing and shoes suitable for exercise. Generally, men and women should be dressed in exercise shorts or pants with a T-shirt or sleeveless blouse. Women should wear a two-piece bathing suit top or exercise sport top under a T-shirt or blouse to permit electrocardiography measures.

d. The Health and Lifestyle History (example included in this laboratory chapter) with instructions to complete it before arrival at the testing facility.

2. The student should master the objectives of Laboratory 6, "Resting and Exercise Electrocardiography" as well as the objectives of Laboratory 7 titled "Measuring Resting and Exercise Blood Pressure" before proceeding. It is assumed in the procedures outlined below that the student has completed both of these laboratory experiences, and has carefully reviewed the procedures for each.

3. The technician must be familiar with the operational controls, commands, and features of the equipment that will be used for these testing procedures.

    a. Check to be sure all equipment is clean and in good working condition, and that the treadmill or cycle ergometer have been recently calibrated. Refer to the Appendix for procedures if calibration is necessary.

4. On the day of testing, check the laboratory environmental conditions to ensure a temperature range of 20°C to 22°C (68°F to 71.6°F) and a relative humidity less than 60%. These environmental conditions must be standardized from day-to-day in any exercise testing facility, since physiologic responses to exercise can be affected by these factors (Myers, et al., 2009; Thompson, Gordon, and Pescatello, 2010).

PRELIMINARY SLM-GXT PROCEDURES

1. It is the right of each subject to be fully informed of the procedures and risks of the SLM-GXT, and voluntarily grant his or her consent before testing procedures begin. This process is formally completed by reading and signing the Informed Consent for Symptom-Limited Maximal Graded Exercise Testing. A sample of an informed consent adapted from the ACSM Guidelines for Exercise Testing and Prescription is included in this laboratory chapter (Thompson, Gordon, and Pescatello, 2010). Contact the appropriate institutional authorities for specific policies or additional requirements for obtaining subject informed consent.

    a. The details of the informed consent should be discussed with the subject, and the subject should be given opportunity to ask and have answered questions about the procedure prior to signing.
    b. *The SLM-GXT should not be performed until the informed consent is signed and witnessed.*

2. Carefully review the Health and Lifestyle History Questionnaire the subject was asked to complete before arriving. If not completed, ask the subject to do so before proceeding.

    a. Note current illnesses, presence of cardiovascular disease risk factors, medical history, medications, and physical activity habits. Verbally confirm the information on this questionnaire with the subject.
    b. Use this information to complete the Demographics and Pre-Exercise Assessment of Cardiovascular Disease Risk, Health, and Exercise Readiness sections of the Graded Exercise Test Report found later in this chapter.

3. **DECISION POINT.** Is physician supervision required? It is generally recommended that maximal GXTs be directly supervised by a physician or conducted by trained health care professionals with a physician in the immediate area for men and women with two or more cardiovascular disease risk

factors (Moderate Risk) or who have documented cardiac, pulmonary, or metabolic disease (High Risk). Otherwise, physician supervision and presence is not required (Myers, et al., 2009; Thompson, Gordon, and Pescatello, 2010).

PRE-EXERCISE SUPINE RESTING PROCEDURES

1. Review and duplicate the procedures for Electrode Placement and Skin Preparation and Recording the Resting Supine 12-Lead Electrocardiogram described in Laboratory 6, "Resting and Exercise Electrocardiography". Be sure the ECG recording is labeled appropriately.
2. **Optional procedure for direct $\dot{V}O_2$ measures:** calibrate the metabolic cart during this pre-exercise period.
3. After the resting, supine ECG has been recorded, measure resting, supine blood pressure following the procedures detailed in Laboratory 7, "Measuring Resting and Exercise Blood Pressure". Record the systolic and diastolic blood pressure values on the resting, supine ECG just previously recorded. Many automated systems permit keyboard blood pressure inputs, which are then printed on the ECG.
4. Record the values for resting heart rate and blood pressure in the Resting ECG and Blood Pressure Data section of the Graded Exercise Test Report.
5. Before proceeding, an interpretation of the resting ECG and blood pressure values should be provided to rule out any risks for heavy exercise. This interpretation will usually be completed by a physician if required for the GXT (refer to item 3 in the Preliminary SLM-GXT Procedures section for physician presence guidelines), or by a person knowledgeable in ECG interpretation.
    a. The student should refer to the Model for ECG Interpretation found in Laboratory 6, "Resting and Exercise Electrocardiography" to aid in interpreting the ECG.
    b. Resting blood pressure should not exceed 200 mmHg systolic or 110 mmHg diastolic to proceed.
    c. Check the response corresponding to the ECG interpretation in the Resting ECG and Blood Pressure Data section of the Graded Exercise Test Report.
6. When physician supervision of the SLM-GXT is necessary, the pre-exercise physical exam can be conducted at this point in the procedure.
    a. The attending physician should examine the recorded resting ECG and blood pressure values, as well as perform a general physical examination covering the items listed in the Physical Examination section of the Graded Exercise Test Report.
    b. In the absence of a physician to conduct the physical exam, this section should be left blank.
7. **DECISION STOP/GO POINT.** This is the committed step of the SLM-GXT. A thorough review of all resting data and health history information should be conducted by the physician (if required) and by the other testing personnel charged with administering the SLM-GXT. Consider this information in light of possible Contraindications to Graded Exercise Testing (Table 9.5).
    a. The SLM-GXT should generally not be conducted if Absolute Contraindications are noted, except under carefully controlled conditions in a clinical environment.
    b. In the case of Relative Contraindications, the benefits of the SLM-GXT must be judged to outweigh the risks of the procedure.
    c. Unless directed by a physician, **do not** conduct the SLM-GXT when in doubt about risks of complications for a subject.
    d. Complete and sign the GXT Authorization section of the Graded Exercise Test Report before proceeding.

TABLE 9.5 – Absolute and relative contraindications to graded exercise testing

ABSOLUTE CONTRAINDICATIONS

1. ECG contraindications
   a. Recent resting ECG change suggesting ischemia, infarction or other acute cardiac problems
   b. Uncontrolled ventricular or atrial arrhythmia causing symptoms or compromising heart function
   c. Third-degree A-V block

2. Unstable angina, uncontrolled symptomatic heart failure, or recent complicated myocardial infarction
3. Severe aortic stenosis
4. Suspected or known dissecting aneurysm
5. Active or suspected myocarditis or pericarditis
6. Thrombophlebitis or intracardiac thrombi
7. Pulmonary embolus or pulmonary infarction
8. Acute infection with fever, lymph gland inflammation, or general body aches
9. Significant emotional distress
RELATIVE CONTRAINDICATIONS
1. Frequent or complex ventricular ectopic beats, bradyarrhythmias or tachyarrhythmias
2. High-degree atrioventricular block
3. Left main coronary artery stenosis
4. Moderate valvular disease
5. Elevated resting blood pressure; systolic > 200 or diastolic > 110 mmHg
6. Known electrolyte disturbances, such as hypokalemia
7. Ventricular aneurysm
8. Heart outflow tract obstruction, such as hypertrophic cardiomyopathy
9. Uncontrolled metabolic diseases, such as diabetes
10. Chronic infectious diseases (e.g., AIDS, hepatitis)
11. Advanced or complicated pregnancy
12. Musculoskeletal or neuromuscular disorders or other physical disorders that could be significantly aggravated by exercise

Adapted from Gibbons et al., 2002; Thompson, Gordon, and Pescatello (eds), 2010.

EXERCISE PROCEDURES

1. With minor modifications, the procedures to follow extend to the SLM-GXT the techniques for measuring the exercise ECG and blood pressure detailed in Laboratories 6 and 7. Optional procedures include direct measures of $\dot{V}O_2$, and use of a stationary cycle ergometer.
2. Review all subject data collected up to this point, including data in the Health and Lifestyle History and from informal subject interviews, for any factors which might influence the subject's exercise response, such as medications or medical history of existing symptoms.

a. Briefly note these factors under the Pre-Exercise Data section of the Graded Exercise Test Worksheet. These notes serve as a reminder during the exercise test of conditions that could be related to the development of symptoms or cause unusual exercise responses.

3. Also consider the all subject data in selecting the appropriate SLM-GXT protocol and modality. A protocol is adopted in most testing laboratories to be applied in a majority of cases. However, exceptions must be considered when subject conditioning or physical ability warrant a protocol change.

    a. Refer to Table 9.6 for the most frequently used continuous, multistage treadmill protocols and a standard protocol for a stationary cycle ergometer. In general, any protocol should include a low-intensity warm-up (two to three MET intensity), with no more than three MET increments each one to three minute stage.
    b. Protocols with smaller work increments should be selected for those who are elderly or obese, and for those in very poor physical condition either as a result of deconditioning or disease.

TABLE 9.6 – Common treadmill and cycle ergometer protocols with estimated Mets per stage

PROTOCOL	Stage	Grade (%)	Speed (mph)	Total Time (min)	Mets
Bruce, Kusumi, and Hosmer, 1973	1	10	1.7	3	4.6
	2	12	2.5	6	7.0
	3	14	3.4	9	10.2
	4	16	4.2	12	12.1
	5	18	5.0	15	14.9
	6	20	5.5	18	17.0
Balke and Ware, 1959	1	0	3.4	1	3.6
	2	2	3.4	2	4.5
	3	3	3.4	3	5.0
	4	4	3.4	4	5.5
	5	5	3.4	5	5.9
	6	6	3.4	6	6.4
	For each additional stage, ↑ grade 1% every minute				
Naughton and Hellerstein, 1973	1	0	1.0	2	1.9
	2	0	2.0	4	2.8
	3	3.5	2.0	6	3.8
	4	7.0	2.0	8	4.7
	5	10.5	2.0	10	5.6
	6	14	2.0	12	6.5
	7	17.5	2.0	14	7.5
	8	12.5	3.0	16	8.4
	For each additional stage, ↑ grade 3.5% every 2 minutes.				
CYCLE ERGOMETER	1 minute warm-up at 50–60 rpm's and workrate of 150 kpm · min^{-1} (25 W). Increase 75 to 150 kpm · min^{-1} (12.5 to 25 W) every minute until exhaustion.				

c. Cycle ergometer or arm ergometer modes of exercise may be chosen for those unable to negotiate treadmill walking.
d. The standard Bruce protocol (Bruce, Kusumi, and Hosmer, 1973) has been chosen for this laboratory exercise. Enter the protocol name in the "Protocol" blank of the Graded Exercise Test Report and also on the Graded Exercise Test Worksheet. Set the computerized ECG controller (if available) to the Bruce protocol. If a different protocol is chosen, make the necessary adjustments in the forms and computer selection.
   i. Write the selected stage times, speed (or rpm), and grade (or kpm) in the appropriate blanks in the Exercise Data section of the Graded Exercise Test Worksheet.
4. An individualized maximal heart rate target for the GXT is necessary to aid the technician in anticipating the voluntary termination of the test. Record on the Graded Exercise Test Worksheet a maximal heart rate (HR_{max}) if known from a previous GXT. If HR_{max} is unknown, several prediction equations exist to obtain an estimate, including a simple estimates as follows (Gellish et al., 2007; Thompson, Gordon, and Pescatello, 2010).

Predicted Maximal Heart Rate = 207 − (0.7 × subject's age in years)

OR = 220 − subject's age in years.

a. Also calculate 85% of this expected HR_{max}, and record. A GXT is not considered maximal if the exercise heart rate does not exceed 85% of the expected maximum value, unless the heart rate is limited by medications, such as beta-adrenergic blocking agents.
5. Attach the ECG lead harness or telemetry system securely to the waist and escort the subject to the treadmill (or cycle ergometer).
6. Thoroughly explain the SLM-GXT testing procedures to the subject.
   a. Describe the exercise intensity changes and physiologic measurements to be expected with each stage.
   b. Stress that the protocol will continue increasing in intensity with each stage until maximal exercise effort is achieved and the subject voluntarily elects to end the test.
      i. **Optional procedures for direct $\dot{V}O_2$ measures.** Explain the procedures for collecting metabolic gases during the test.
   c. Specify that the subject should use the word "STOP" clearly spoken (or "thumbs down" sign) when he or she wishes to end the test, and he or she should simultaneously grasp the handrails to maintain balance as the treadmill speed slows, and the elevation returns to zero. (For the cycle ergometer, the resistance will be reduced.)
   d. Make clear that there will be a walking cool-down after maximal effort is achieved, and that the treadmill will not abruptly stop when the subject requests to end the test.
   e. Explain the use and interpretation of the Rating of Perceived Exertion (RPE) scale developed by Borg (1982). Most GXT laboratories adopt the consistent use of either the 10 or 15 grade scale, and although both are valid, the two cannot be used interchangeably (Table 9.7).

7. Demonstrate to the subject a safe way to step onto the moving treadmill (alternately, pedaling a cycle ergometer). This is usually done by the blood pressure technician, who has the responsibility of maintaining communication with the subject during the test.
   a. First, step onto the unmoving treadmill belt, gasp the safety rails, and straddle the treadmill belt.
   b. Ask the ECG technician to start the belt at 1.5 mph and 0% grade.
   c. From the straddle position, lift one leg and gently "paw" the belt to get the feel of the belt speed.
   d. Keep holding the safety rails, step on the moving belt, and begin walking. Once comfortable with the speed, release the hands from the safety rails and begin walking normally on the moving belt.
   e. Step off the treadmill and turn off the treadmill belt.
8. Complete steps to measure the pre-exercise ECG and blood pressure.
   a. Ask the subject to step onto the treadmill and stand quietly on the unmoving treadmill belt (sit, if cycle ergometer) with arms relaxed at the side.
   b. Toggle the 12-lead switch or computer key to record the standing (seated if cycle ergometer) resting, pre-exercise 12-lead ECG.
      i. Review the recorded 12-lead, and if excess artifact is present repeat the recording. Do not proceed to exercise until a high-quality pre-exercise ECG is recorded.
      ii. Label this ECG as STANDING REST (or SEATED if cycle ergometer). Label this and all subsequent ECG recordings, by hand if not automated by the ECG machine, with the subject's name, date, exercise stage, exercise time, and workload (i.e., speed and grade of the treadmill or rpm and watts of the cycle ergometer).
   c. Measure the pre-exercise standing (seated if cycle ergometer) blood pressure, but not until after the ECG has been recorded so as not to add movement artifact to the ECG recording.
      i. Do not remove the blood pressure cuff from the subject's arm. The cuff may be taped to the upper arm to help keep it in place during exercise.
   d. Record the values for pre-exercise heart rate and blood pressure in the Pre-Exercise Data section of the Graded Exercise Test Worksheet.
   e. **Optional procedures for direct $\dot{V}O_2$ measures**. Attach the mouthpiece or mask and nose clip to the subject. Activate the computer key or control switch of the metabolic cart to begin collecting resting metabolic gases.
      i. Continue resting collection for at least one minute. Confirm on the computer monitor that the metabolic cart is collecting resting data, and that the values are reasonable for standing at rest. (Recall that one MET is $3.5 \, ml \cdot kg^{-1} \cdot min^{-1}$.)
9. Initiate the SLM-GXT procedure.
   a. Instruct the subject to grasp the safety handrails, and straddle the unmoving treadmill belt as demonstrated previously.
   b. Carefully check to be sure the subject is completely off the treadmill belt, and start the belt at 0% grade and 1.5 mph. Coach the subject through the safe method to step on the moving treadmill and walk.
      i. When testing older and more unstable subjects the technician should place a hand on the back of the subject to protect against an accidental fall when beginning to walk on the treadmill.

TABLE 9.7 – The 15- and 10-grade rating of perceived exertions scale

15-Grade Scale		10-Grade Scale	
6		0	Nothing at all
7	Very, very light	0.5	Very, very weak (just noticeable)
8		1	Very weak
9	Very light	2	Weak (light)
10		3	Moderate
11	Fairly light	4	Somewhat strong
12		5	Strong (heavy)
13	Somewhat hard	6	
14		7	Very strong
15	Hard	8	
16		9	
17	Very hard	10	Very, very strong (almost max)
18			
19	Very, very hard		
20			Maximal

From Borg, 1982.

    c. As soon as the subject is comfortably walking, ask him or her to release the support rails and walk naturally.

        i. Holding the handrails during the SLM-GXT should be discouraged, unless absolutely necessary for the subject to maintain balance, and then a light touch on the rails is usually all that is required. Allowing the subject to hold or pull on the handrails will spuriously prolong the test and result in an overestimation of oxygen uptake predicted from published equations using time on the treadmill, or the speed and grade at maximal exercise. Measured $\dot{V}O_{2max}$ negates this problem.

    d. Verbally alert the subject that the exercise test is about to begin, then toggle the switch or computer key on the ECG/treadmill machine to initiate Stage 1 of the selected protocol (refer to Table 9.6).

        i. **Optional procedure for direct $\dot{V}O_2$ measures**. Coordinate the start of Stage 1 on the ECG machine with activating the control key or switch on the metabolic cart to initiate collection of metabolic gases during exercise. See Figure 9.1(*a*) and 9.1(*b*) for a typical setup to measure $\dot{V}O_2$ using an automated system.

        ii. **Optional cycle ergometer**. The same general sequence of procedures should be followed, except the mode of exercise will be a stationary cycle ergometer.

FIGURE 9.1(a) – Typical configuration for a SLM-GXT on a treadmill with measured oxygen uptake.

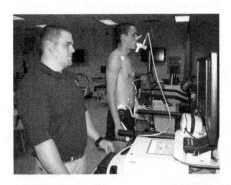

FIGURE 9.1(b) – A screen shot with the subject at standing rest of monitors from a stress testing system with the capability of measuring both the ECG and metabolic gases during a SLM-GXT.

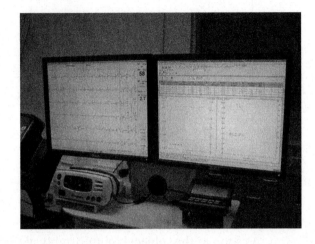

10. General responsibilities of the <u>blood pressure technician</u> during the SLM-GXT.
    a. Within the first minute of the start of the test, check the blood pressure cuff to ensure that it is properly positioned for blood pressure measurements.
    b. Follow the general procedures for measuring exercise blood pressure during each exercise stage as detailed in Laboratory 7. Refer to Figure 9.2 for the typical arrangement of the technician as he or she takes the exercise blood pressure. As a brief review,
        i. Acquire at least one blood pressure reading each stage, usually during the last minute of the stage. Start early enough to be finished *before* the ECG recording is initiated the last 10–12 seconds of each stage. Recording the ECG usually takes precedence over measuring the blood pressure.
        ii. Be sure to support the subject's fully extended arm at about heart level, and place the bell of the stethoscope flat on the skin directly over the brachial artery in the antecubital fossa.
        iii. Systolic blood pressure will rise during exercise, so the cuff inflation pressure will need to be higher than at rest.
        iv. After the measurement is made, verbally report the reading to the ECG technician who will record it on the Graded Exercise Test Worksheet.
        v. Attempt to time the final blood pressure measurement so that it is made as closely as possible to maximum exercise. This will require close observation of the subject for cues that the subject is nearing maximal effort.
            1. Heart rate (HR) above 85% and a high RPE are cues the subject is nearing maximal effort.

FIGURE 9.2 – The typical setup for taking blood pressures during a treadmill SLM-GXT. Note the correct technique of the technician supporting the subject's arm at about heart level with the elbow at complete extension, the bell of the stethoscope firmly in place over the location of the artery, and the subject's hand grasping the upper arm of the technician. Not visible, the technician is elevated by standing on a step box or stool.

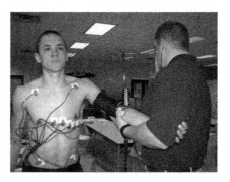

    c. Maintain communication with the subject.
        i. Ask for an RPE value corresponding to the subject's perceived exertion during the last 15 seconds of each stage, and report the value to the ECG technician who will record it.
        ii. As symptoms of maximal exertion become evident, encourage the subject to continue as long as possible, and remind the subject to verbally say the word "STOP" (or give "thumbs down" signal) and grasp the handrails when he or she is too fatigued to go on.
        iii. Remember, the GXT should be stopped for any of the reasons shown in Table 9.1.
    d. **Optional procedure for direct $\dot{V}O_2$ measures.** Since the subject is breathing through a mouthpiece or mask and unable to talk, hold a copy of the RPE scale in front of the subject, and ask the subject to point to the RPE number corresponding to his or her perceived exertion.
        i. Instruct the subject to give a "thumbs-up" signal to continue and to give a "thumbs-down" signal and grasp the handrails when he or she wishes to stop.
11. General responsibilities of the ECG technician during the GXT.
    a. The primary responsibility is to monitor the ECG for cardiac problems during exercise, and terminate the test when the subject requests to do so or when a Reason for Stopping a SLM-GXT (Table 9.1) is noted. All other responsibilities are secondary.
    b. Immediately correct the problem if the ECG becomes unreadable at any time for any reason, and stop the test if necessary.
    c. Ensure that an interpretable ECG, usually a 12-lead, is recorded at the following times during the GXT. (Computerized ECG equipment is usually programmed for the desired ECG frequency during exercise and recovery.) Immediately repeat a recording if the ECG is uninterpretable.
        i. The last 10–12 seconds of every stage completed.
        ii. When the heart rate reaches 85% of predicted maximum.
        iii. At maximum exercise as part of the END EXERCISE protocol (see below).
        iv. Immediately post-exercise (IPE).
        v. At least every 2 minutes of recovery.
    d. Manually record the ECG any time an abnormal ECG waveform or rhythm is observed to capture the anomaly for future reference. Note this in the Comments section of the Graded Exercise Test Worksheet.

e. Be sure that each ECG printed is labeled with the subject's name, the date, and the blood pressure corresponding to that stage. Often these entries are automated by the ECG equipment, if so programmed.
f. Record GXT data in the Exercise Data section of the Graded Exercise Test Worksheet.
   i. The heart rate (HR) is usually calculated automatically by the ECG equipment. However, the accuracy of these calculations should always be verified by the technician, especially when an excess amount of artifact is present in the ECG signal, which often is the case near maximal exertion.
   ii. The blood pressure (BP) and RPE readings should be recorded on the Graded Exercise Test Worksheet when reported by the blood pressure tech.
   iii. Enter any relevant comments in the Comments section of the Graded Exercise Test Worksheet about the time and severity of any symptoms whether normal fatigue symptoms (e.g., leg fatigue) or abnormal symptoms, such as chest pain.
   iv. **Optional procedures for direct $\dot{V}O_2$ measures**. Record the $\dot{V}O_2$ values given by the metabolic cart at the end of each stage on the Graded Exercise Test Worksheet.
g. The responsibility of starting and stopping the GXT and controlling the stage increments rests with the ECG tech.

SLM-GXT TERMINATION AND RECOVERY

1. The ECG tech should activate the appropriate equipment control switch or computer key to initiate the END EXERCISE protocol immediately on the STOP ("thumbs down") command from the subject, or when noting any one of the Reasons for Stopping a SLM-GXT given in Table 9.1.
   a. The END EXERCISE protocol for a Functional SLM-GXT consists of simultaneously recording a 12-lead ECG labeled as *MAX*, and reducing the treadmill speed to 1.5 to 2 mph and the grade to 0% (decreasing resistance to 0.25 or 0.5 kp on a cycle ergometer) to initiate a post-exercise active cool-down protocol. On most computer-controlled ECG exercise test systems, this sequence will be engaged automatically if appropriately programmed into the system.
      i. **Optional procedure for direct $\dot{V}O_2$ measures**. Coordinate activating the ECG/treadmill system END EXERCISE sequence with the same on the metabolic cart. Most metabolic carts will continue to sample metabolic gases during recovery unless commanded to stop collecting data, but will mark the END EXERCISE time in the data file, which can be printed and reviewed at a later time.
   b. Record the total amount of exercise time in minutes and seconds the subject completed on the exercise protocol. This is usually given on the ECG monitor and recorded automatically on the MAX ECG printout. Manually write this time on the MAX ECG printout if not done so automatically by the ECG system.
      i. Also record this as Max Time in the Maximal Exercise Summary portion of the Graded Exercise Test Worksheet.
   c. **Alternate post-exercise protocol to maximize diagnostic sensitivity for a Diagnostic SLM-GXT** (Thompson, Gordon, and Pescatello, 2010). In place of an active cool-down, the subject should assume a supine position during the recovery period. This is typically not done for a Functional SLM-GXT.
   d. **Stop the treadmill, abort the active cool-down, and place the subject in the supine position at any time emergency dictates.**

e. At no time should the subject be left to stand immediately after maximum exercise, since hypotension, dizziness, and fainting may ensue.
      i. Post-exercise hypotension and symptomatic dizziness and lightheadedness can usually be treated by placing the subject in the supine position and elevating the subject's legs about 12 to 18 inches by placing a support under the feet.
   f. The blood pressure technician should obtain a blood pressure as soon as possible after the END EXERCISE protocol is initiated. Because blood pressure may be difficult to obtain during the final stages of exercise, especially if the subject is running, this blood pressure may be considered maximal blood pressure.
2. Cool-down protocol.
   a. A second ECG, labeled *IPE*, should be recorded as soon as the subject reaches the cool-down workload (or supine position), usually within the first 30 seconds of initiating the cool-down protocol. Often the ECG recorded at maximum exertion is difficult to interpret due to motion artifact. The IPE ECG may, therefore, assume a critical place of importance in interpreting the overall results of the SLM-GXT.
   b. The blood pressure and ECG should be measured and recorded every two minutes for at least 4–6 minutes of recovery, or until the heart rate is below 100, blood pressure has normalized, and symptoms, if present, have dissipated.
      i. Early in recovery the blood pressure technician should query the subject to determine his or her final RPE at voluntary maximal effort, and ask for a subjective reason for stopping exercise.
      ii. Pay close attention to all vital signs and the ECG during recovery. Abnormalities that continue from the exercise test or are newly developed during recovery are known to have important diagnostic and prognostic relevance (Myers, et al., 2009).
   c. **Optional procedure for direct $\dot{V}O_2$ measures**. Collect metabolic gases for one to two minutes of recovery, and then switch the control on the metabolic cart to end data collection and remove the subject's mask or mouthpiece.
   d. Record all recovery data and important comments in the Recovery Data portion of the Graded Exercise Test Worksheet.
3. When the subject has completed the cool down, verbally alert him or her, then activate the equipment switch or computer key to stop the treadmill belt.
   a. If cycle ergometer test, instruct the subject to stop pedaling.
4. Unless instructed by a physician or other knowledgeable laboratory personnel to continue monitoring the subject, the subject can now be disconnected from the monitoring equipment and released from the testing area.
   a. The subject may be held in the testing area for extended observation if there is any question as to his or her complete recovery from exercise.

POST-EXERCISE DATA REDUCTION AND REPORT PROCEDURES

1. The ECG technician should enter into the Maximal Exercise Summary portion of the Graded Exercise Test Worksheet the exercise *Max Time*, *Max HR*, and *Max BP* as well as the *Reason for Stopping* (Stop Reason).
   a. Verify all data are correct on the Graded Exercise Test Worksheet before proceeding to develop the SLM-GXT report.

b. Check the heart rate values recorded on the worksheet against those derived from a visual inspection of the ECG at each stage. Be aware the automated calculations may not be accurate, especially if during exercise the ECG was distorted by artifact. Standardize the heart rate calculations by always using the first 6 seconds of each ECG recording.

2. Complete the Maximal Exercise Values section of the Graded Exercise Test Report.

   a. Transfer the necessary values from the Graded Exercise Test Worksheet to the Pre-Exercise and Maximal Exercise Values section of the Graded Exercise Test Report. Data to be transferred include: (1) exercise protocol used, (2) pre-exercise (Preex) heart rate (HR) and blood pressure (BP), (3) maximum HR (MHR) and BP (MBP) achieved on the SLM-GXT, (4) RPE at maximum exercise, (5) treadmill speed and grade (or Watts if cycle ergometer), (6) max Time in minutes:seconds, and (7) Reason for Stopping.

   b. Calculate the percent of maximal heart rate that was achieved on the SLM-GXT (%Pred Max), then enter the value on the report form.

   $$\%\text{Pred Max} = (\text{MHR achieved} \div \text{Predicted Max HR}) \times 100$$

   Where: MHR achieved = maximal HR recorded on the GXT

   Predicted Max HR = calculated in Item #4 of Exercise Procedures

   c. Calculate the rate pressure product (RPP) at maximal exercise, and enter the value on the report form.

   $$\text{RPP} = (\text{MHR achieved} \times \text{Maximal systolic blood pressure}) \div 100$$

   d. Calculate the subject's achieved $\dot{V}O_{2max}$ (ml·kg^{-1}·min^{-1}) using the treadmill-specific equation for the Bruce protocol from Table 9.2. Note that in these equations maximal exercise "time" must be expressed as minutes and the decimal equivalent of seconds (e.g., 9 minutes and 30 seconds = 9.5 minutes).

      i. **Optional procedure for direct $\dot{V}O_2$ measures.** Enter the directly measured $\dot{V}O_{2max}$ value in the place of the calculated value.

   e. METS are often used in prescribing exercise or in communicating maximum working capacity information in clinical settings. Calculate maximum working capacity in MET values as:

   $$\text{METS} = \dot{V}O_{2max} \text{ (ml·kg}^{-1}\text{·min}^{-1}\text{)} \div 3.5 \text{ (ml·kg}^{-1}\text{·min}^{-1}\text{)}$$

   f. Calculate Functional Aerobic Impairment (FAI). It is useful to compare a subject's achieved $\dot{V}O_{2max}$ to some normative value, and in this way gain insight into whether the subject is average, below average, or above average in aerobic endurance capacity. The FAI is one method to calculate a normative comparison by expressing an individual's $\dot{V}O_{2max}$ as a percentage of one's age- and gender-specific predicted value above or below average (Bruce, Kusumi, and Hosmer, 1973). Calculating the FAI is a two-step process.

   First, calculate predicted $\dot{V}O_{2max}$ (ml·kg^{-1}·min^{-1}) using the appropriate regression equation:

   Sedentary Men: Predicted $\dot{V}O_{2max} = 57.8 - (0.445 \times \text{age in years})$

   Sedentary Women: Predicted $\dot{V}O_{2max} = 42.3 - (0.356 \times \text{age in years})$

   Second, calculate the FAI as follows:

   $$\text{FAI} = [(\text{Predicted } \dot{V}O_{2max} - \text{achieved } \dot{V}O_{2max}) \div \text{Predicted } \dot{V}O_{2max}] \times 100$$

Note that a <u>negative</u> FAI indicates an individual is **above** average in aerobic endurance capacity, while by comparison a positive FAI indicates an individual is **below** average.

g. Using the Exercise Data from the Graded Exercise Test Worksheet, complete the Exercise Response Chart by graphing heart rate, systolic, and diastolic blood pressure values (Y-axis) measured at the *time in minutes* corresponding to the end of each GXT stage (X-axis). Graph the pre-exercise values at minute zero (0). Connect the points to construct a line graph showing the data trends for each specific variable.

3. Complete the Exercise Interpretation, Recovery Interpretation and Interpretation Summary and Recommendation portions of the Graded Exercise Test Report. The attending physician will generally review and complete these portions of the report. When physician presence is not required by policy (Myers, et al., 2009; Thompson, Gordon, and Pescatello, 2010), others qualified and experienced in exercise testing may complete these portions of the report. See Tables 9.1, 9.4, and 9.8 for information that will be helpful in interpreting the GXT results.

a. Review the Exercise Response Chart and the data reported in the Pre-Exercise and Maximal Exercise Values section to interpret the heart rate and blood pressure responses. If the physiologic responses are deemed "abnormal" indicate the nature of this conclusion by checking the appropriate box in the Interpretation Key at the bottom of the Graded Exercise Test Report.

TABLE 9.8 – Possible exercise responses to consider in interpreting the GXT

NORMAL OR NOT INDEPENDENTLY DIAGNOSTIC
ECG
1. Shortened Q-T interval. 2. Reduced R-wave amplitude, especially over the left ventricle. 3. J-point displaced below isoelectric line. 4. Positive upslope on the ST-segment returning to isoelectric by .08 sec past J-point. 5. Isolated or occasional superventricular arrhythmias, including short runs of SVT. 6. Isolated or occasional ventricular ectopic beats.
Physiologic
1. Linear increase in HR with increased workload averaging 10 ± 2 beats $\cdot$ MET^{-1}, and a max HR within $\pm 15\%$ of predicted max. 2. Decrease in HR during recovery (HR-recovery) averages > 12 beats $\cdot$ min^{-1} in first minute of recovery and > 22 beats $\cdot$ min^{-1} by end of second minute. 3. A nearly linear increase in systolic BP of 10 ± 2 mmHg $\cdot$ MET^{-1}, reaching a maximal value $\geq$ 140 mmHg but <250 mmHg at maximal exertion. 4. No change or a slight decrease in diastolic BP with exercise. 5. Average to above average aerobic capacity measured by $\dot{V}O_{2max}$, FAI ≤ 0.
Symptoms
1. Exercise limited by fatigue, either general or specific to limb muscles. 2. RPE ≥ 17 (very hard). 3. Normal shortness of breath (hyperpnea).

Continued

ABNORMAL
ECG
1. Horizontal or downsloping ST-segment depression ≥1 mm (≥.1 mv) extending for .08 sec past the J-point.
2. ST-segment elevation ≥1 mm (≥.1 mv).
3. Increased R-wave amplitude with exercise.
4. Sustained supraventricular tachycardia.
5. Ventricular tachycardia.
6. Appearance of complex ventricular ectopy; short runs of ventricular tachycardia, multiform PVCs, couplets.
Physiologic
1. HR at max > 2 standard deviations below age-predicted max or inability to achieve 85% of predicted max HR. Termed chronotropic incompetence.
2. Delayed decrease in HR during recovery (HR-recovery) defined as <12 beats · min^{-1} in first minute of recovery and <22 beats · min^{-1} by end of second minute.
3. Exertional hypotension defined as:
a. A failure of systolic BP to increase appropriately 10 ± 2 mmHg MET^{-1} with an increase in exercise.
b. A drop of ≥10 mmHg in systolic BP as workload increases.
c. A failure of systolic BP to exceed 140 mmHg at max.
4. Systolic BP ≥250 mmHg, recommended end point for exercise.
5. An increase in diastolic BP with exercise, or diastolic BP >115 mmHg. Exercise should be stopped if diastolic BP >115 mmHg.
6. Below average aerobic capacity as measured by $\dot{V}O_{2max}$, FAI >0.
Symptoms
1. Angina or severe chest pain.
2. Severe shortness of breath (dyspnea), especially at low intensity exercise.
3. Fainting, pallor, cyanosis, extreme dizziness or lightheadedness.

Adapted from Thompson, Gordon, and Pescatello (eds), 2010.

    b. Carefully review the entire ECG record made during pre-exercise, exercise, and recovery of the SLM-GXT procedure. For the ECG novice, the information found in the Model for ECG Interpretation in Laboratory 6, "Resting and Exercise Electrocardiography" will be quite helpful. Also, a review of the normal SLM-GXT ECGs in Figures 9.3(a) through 9.3(g) at the end of this chapter will be useful by comparison. If the ECG is interpreted as anything other than normal, indicate the problem in the Interpretation Key.
    c. Indicate any abnormal symptoms noted during the GXT in the Interpretation Key.

d. In the Interpretation Summary and Recommendation section, provide an overall summary interpretation of the SLM-GXT and make comments relative to factors that may have influenced the GXT outcome.

   i. Check the appropriate recommendation regarding the advisability of the subject beginning or continuing a program of regular physical exercise.
   ii. Sign and date the Graded Exercise Test Report.

QUESTIONS AND ACTIVITIES

1. Using the data from this study, describe the response of heart rate, systolic blood pressure, and diastolic blood pressure to incremental exercise.
   a. Do you consider these responses normal? Why or why not?
   b. Provide physiologic reasons for these responses.
2. A heart attack usually results in the death of a portion of the heart muscle. How would you hypothesize that this may affect heart rate, systolic blood pressure, rate pressure product, and $\dot{V}O_{2max}$ measured during the GXT?
3. Describe the general type of GXT protocol you would choose to test the following subjects. Also tell whether or not a physician should be present during the test.
   a. An 82-year-old female who has been sedentary, but without any limiting symptoms or other known risk factors.
   b. A 22-year-old cross-country runner who complains of occasional dizzy spells during fast-paced running.
   c. An obese male, 37 years old, with an orthopedic knee problem that makes walking painful.
   d. A 52-year-old male who complains of frequent and severe chest pains not associated with any type of physical exertion.
4. A 22-year-old woman weighing 135 pounds completes 15 minutes on the Balke protocol. A second woman, who is 33 years old and weighs 115 pounds, completes 10 minutes on the Bruce protocol. Who is in the best aerobic condition? How do you know?

QUESTIONS AND ACTIVITIES FOR GOING FURTHER

1. Give physiologic evidence for or against the assertion that $\dot{V}O_{2max}$ is primarily limited by cardiac output.
2. A trained female cyclist has a $\dot{V}O_{2max}$ of 66 ml · kg^{-1} · min^{-1} at sea level. If she is then tested at an elevation of 5,000 feet, should she expect a change in her $\dot{V}O_{2max}$? Why or why not?
3. Show graphically and explain the comparative responses you would predict from a SLM-GXT conducted before and after 6 months of aerobic training by a 43-year-old healthy man.
4. If you used a metabolic cart to measure $\dot{V}O_{2max}$, use these data to compare the calculated oxygen uptake to the measured for your team members or class members. If significantly different, give reasons why.

FIGURE 9.3(a) – SLM-GXT ECG 1 Supine.

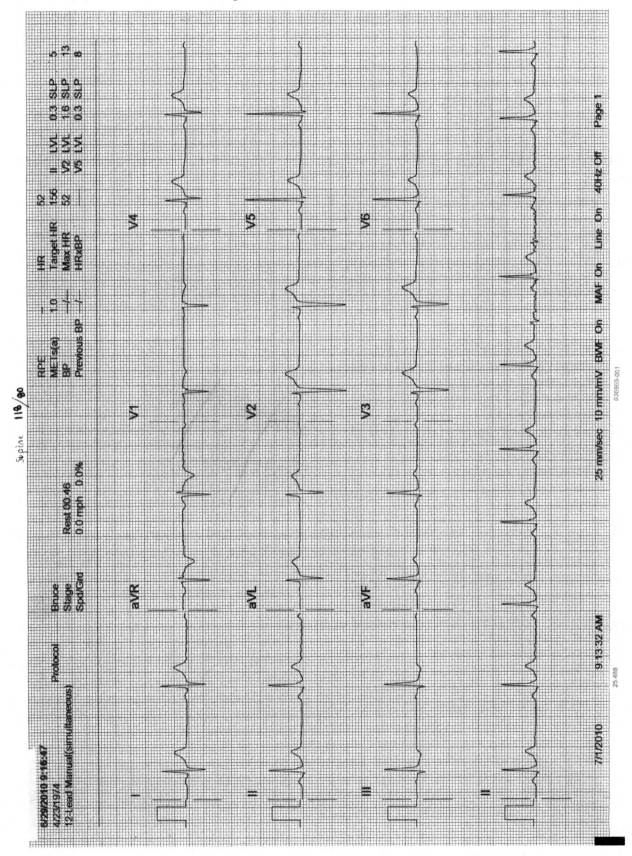

FIGURE 9.3(b) – SLM-GXT ECG 2 Stage 1.

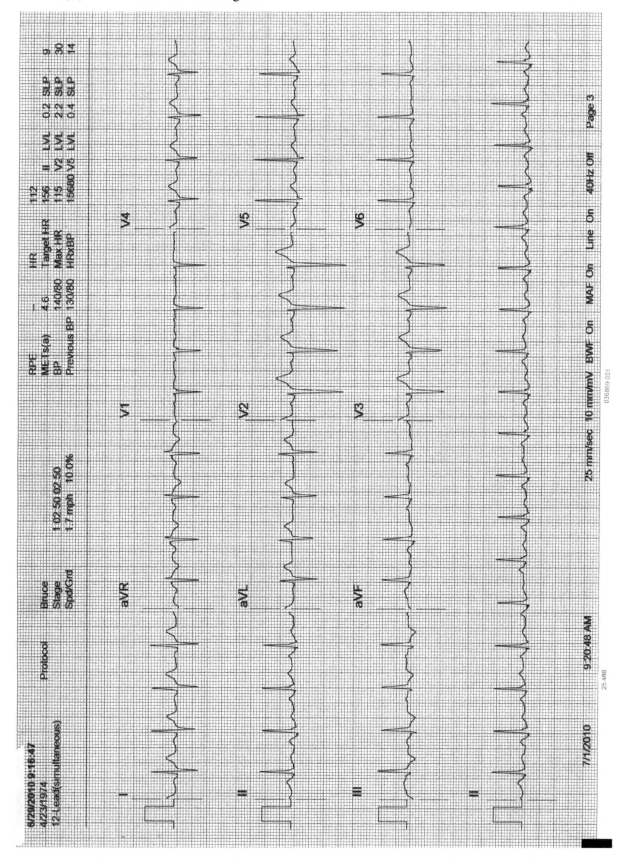

FIGURE 9.3(c) – SLM-GXT ECG 3 Stage 2.

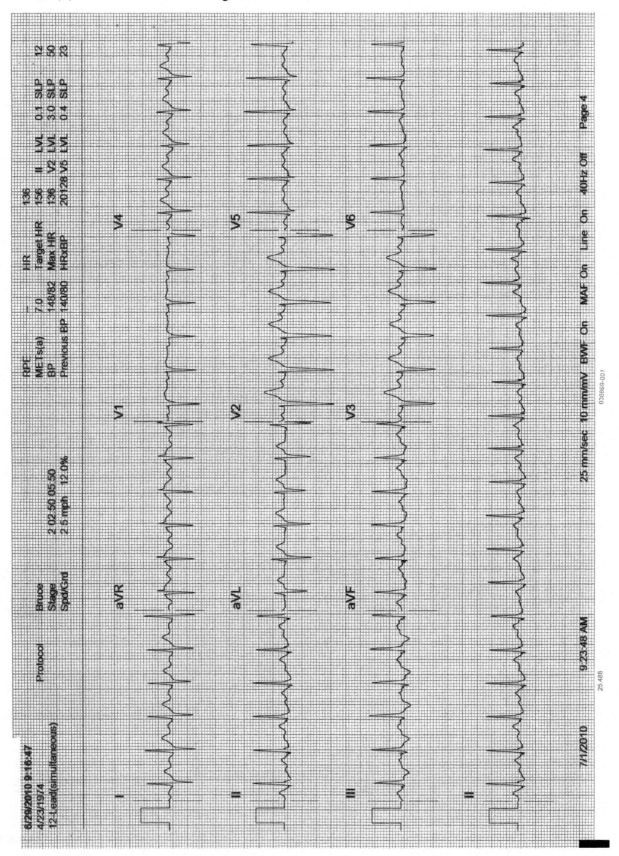

FIGURE 9.3(d) – SLM-GXT ECG 4 Stage 3.

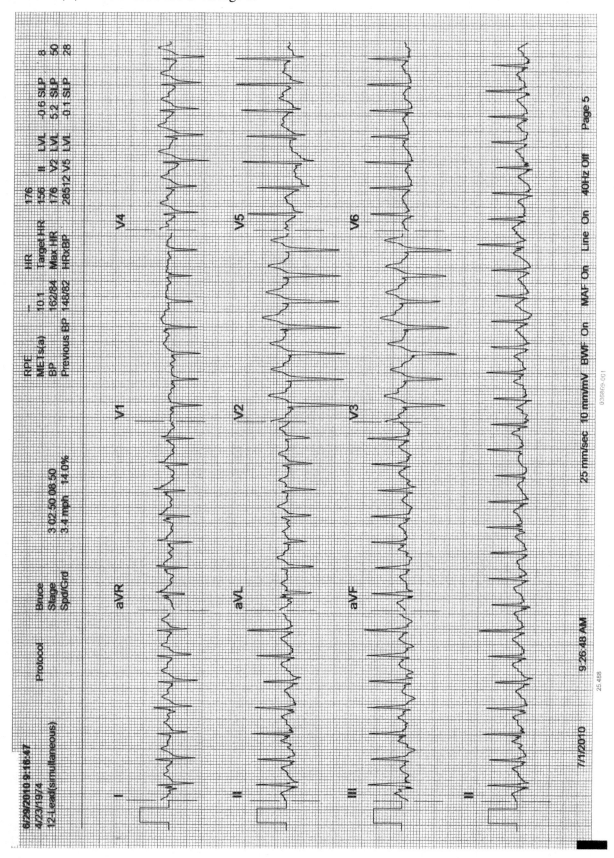

166 The Symptom-Limited Maximal Graded Exercise Test for Clinical and Sports Medicine Applications

FIGURE 9.3(e) – SLM-GXT ECG 5 Max.

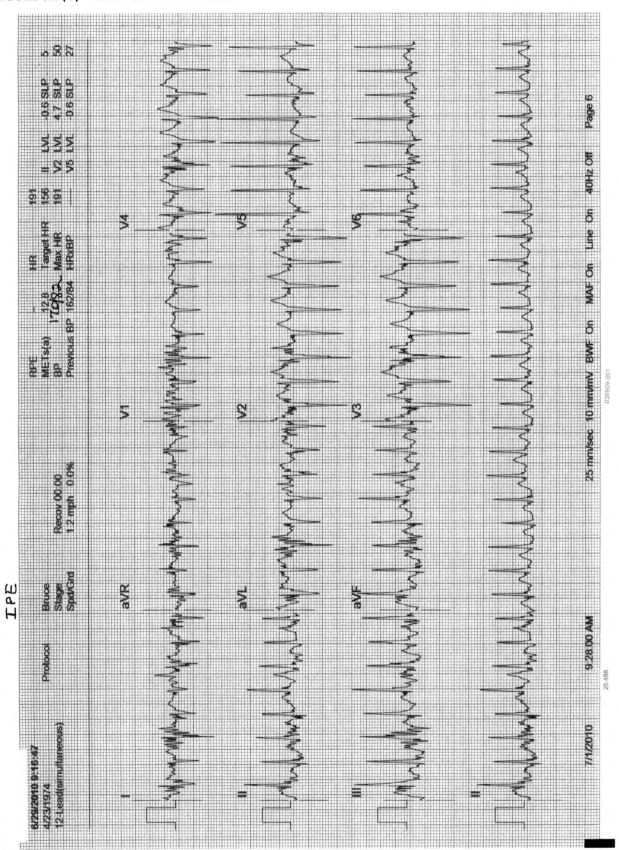

FIGURE 9.3(f) – SLM-GXT ECG 6 IPE 19 sec.

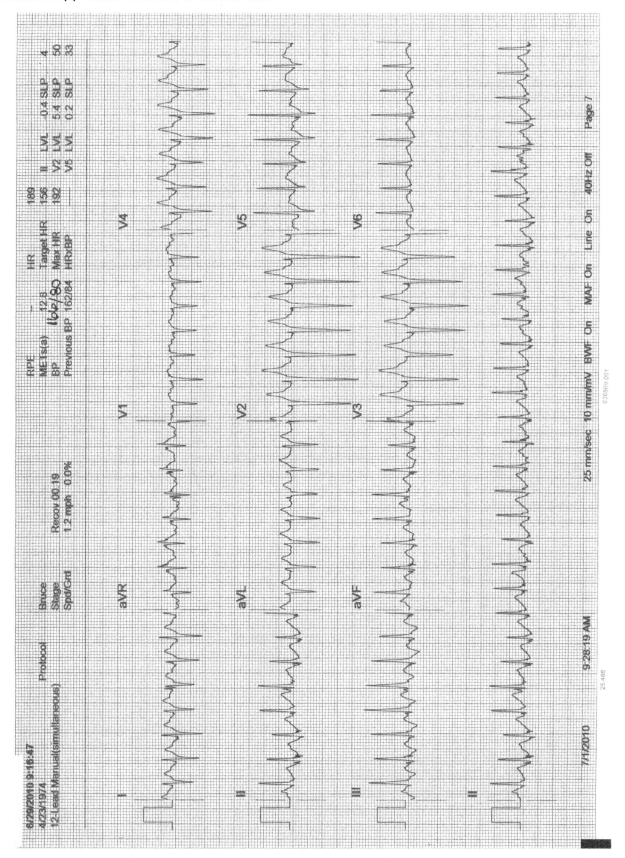

FIGURE 9.3(g) – SLM-GXT ECG 7 Rec 3 min.

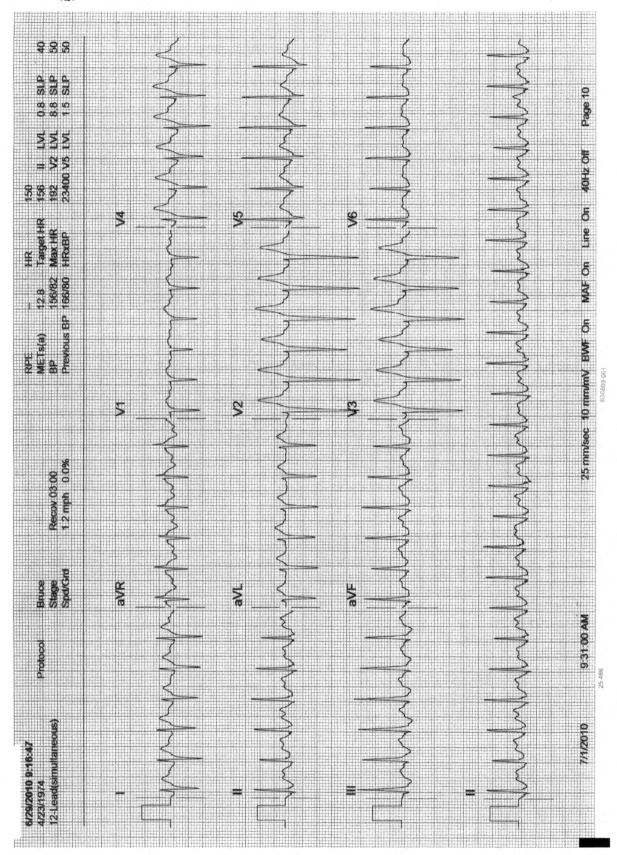

REFERENCES

Balke, B. and R. W. Ware. "An Experimental Study of Physical Fitness of Air Force Personnel." *U.S. Armed Forces Med. J.* 10 (1959): 675–88.

Borg, G. A. V. "Psychophysical Bases of Perceived Exertion." Med.Sci.Sports.Exerc. 14 (1982): 5, 377–81.

Bruce, R. A., F. Kusumi, and D. Hosmer. "Maximal Oxygen Intake and Nomographic Assessment of Functional Aerobic Impairment in Cardiovascular Disease." *Am. Heart J.* 85 (1973): 4, 546–62.

Foster, C., A. S. Jackson, M. L. Pollock, M. M. Taylor, J. Hare, S. M. Sennett, J. L. Rod, M. Sarwar, and D. H. Schmidt. "Generalized Equations for Predicting Functional-Capacity from Treadmill Performance." *Am. Heart J.* 107 (1984): 6, 1229–34.

Foster, C., M. L. Pollock, J. L. Rod, D. S. Dymond, G. Wible, and D. H. Schmidt. "Evaluation of Functional-Capacity during Exercise Radionuclide Angiography." *Cardiology* 70 (1983): 2, 85–93.

Gellish, R. L., B. R. Goslin, R. E. Olson, A. McDonald, G. D. Russi, and V. K. Moudgil. "Longitudinal Modeling of the Relationship between Age and Maximal Heart Rate." *Med. Sci. Sports Exerc.* 39 (2007): 5, 822—29.

Gibbons, R. J., G. J. Balady, J. W. Beasley, J. T. Bricker, W. F. C. Duvernoy, V. F. Froelicher, D. B. Mark, T. H. Marwick, B. D. McCallister, P. D. Thompson, W. L. Winters, and F. G. Yanowitz. "ACC/AHA Guidelines for Exercise Testing: A Report of the American College of Cardiology American Heart Association Task Force on Practice Guidelines (committee on exercise testing)." *J. Am. Coll. Cardiol.* 30 (1997): 1, 260–311.

Gibbons, R. J., G. J. Balady, J. T. Bricker, B. R. Chaitman, G. F. Fletcher, V.F. Froelicher, D.B. Mark, B.D. McCallister, A. N. Mooss, M. G. O'Reilly, W. L. Winters, E. M. Antman, J. S. Alpert, D. P. Faxon, V. Fuster, G. Gregoratos, L. F. Hiratzka, A. K. Jacobs, R. O. Russell, and S. C. Smith. ACC/AHA 2002 Guideline Update for Exercise Testing: Summary Article—A Report of the American College of Cardiology/American Heart Association Task Force on Practice Guidelines (Committee to Update the 1997 Exercise Testing Guidelines). *Circulation* 106 (2002): 14, 1883–92.

Myers, J., R. Arena, B. Franklin, I. Pina, W. E. Kraus, K. McInnis, G. J. Balady, Council Clin Cardiology and Council Cardiovasc Nursing. Recommendations for Clinical Exercise Laboratories: A Scientific Statement from the American Heart Association. *Circulation* 119 (2009): 24, 3144–61.

Naughton, J. and H. K. Hellerstein. Exercise Testing and *Exercise Training in Coronary Heart Disease*. New York: Academic Press, 1973.

Pollock, M. L., R. L. Bohannon, K. H. Cooper, J. J. Ayres, A. Ward, S. R. White, and A. C Linnerud. Comparative Analysis of Four Protocols for Maximal Treadmill Stress Testing. *Am. Heart J.* 92 (1976.): 1, 39–46.

Pollock, M. L., C. Foster, D. Schmidt, C. Hellman, A. C. Linnerud, and A. Ward. "Comparative-Analysis of Physiologic Responses to Three Different Maximal Graded-Exercise Test Protocols in Healthy Women." *Am. Heart J.* 103 (1982): 3, 363–73.

Thompson, W. R., N. F. Gordon, and L. S. Pescatello (eds.). *ACSM's Guidelines for Exercise Testing and Prescription*. Philadelphia: Lippincott Williams & Wilkins, 2010.

HEALTH AND LIFE-STYLE HISTORY

Please complete this form as accurately and completely as possible. The information you provide will be used to evaluate your health by the physician or exercise physiologist who will see you in our laboratory. All information will be treated as privileged and confidential.

IDENTIFICATION AND GENERAL INFORMATION

Name: _____   Today's Date: ____/____/____
      Last   First   M.I.   M/D/Y

Age _____ yr   Date of Birth _____   Sex (M, F) _____

Home Address _____
        Street   City   State   Zip

Office Address _____

Phone: Home _____   Office _____   Cell _____   Occupation _____

Personal Physician _____
        Name   Street   City   State   Zip

ILLNESSES AND MEDICAL PROBLEMS

Check all the conditions or diseases for which you have been diagnosed and/or treated. Also give the date of occurrence or diagnosis. If you suspect that you may suffer from one of the conditions, please indicate this in the right margin after the date.

Condition Diagnosed	Yes	Date (mo/yr)
AIDS	___	_____
Alcoholism	___	_____
Anemia	___	_____
Arthritis	___	_____
Asthma	___	_____
Bronchitis (chronic)	___	_____

Cancer:
  Breast
  Cervix
  Colon
  Lung
  Uterus
  Other _____
Cirrhosis (liver)
Colitis (ulcerative)
Depression
Diabetes
Emphysema
Epilepsy
Frequent Bleeding
Hepatitis B
Pneumonia
Tuberculosis
Renal/Kidney Problems
Other _____

Cardiovascular Problems Diagnosed

	Yes	Date (mo/yr)
Stroke	___	_____
Heart Attack	___	_____
Coronary Disease	___	_____
Rheumatic Fever	___	_____
Rheumatic Heart Disease	___	_____
Heart Valve Problem	___	_____
Heart Murmur	___	_____
Enlarged Heart	___	_____
Heart Rhythm Problem	___	_____
Other Heart Problems	___	_____
High Blood Pressure (controlled)	___	_____
High Blood Pressure (uncontrolled)	___	_____
High Blood Cholesterol	___	_____
Diseases of the Arteries	___	_____
Phlebitis	___	_____
Systemic or Pulmonary Embolus	___	_____
Other _____	___	_____

Do You Now Have or Have You Recently Had:

	Yes	Most Recent Occurrence (mo/yr)
Seizures	___	_____
Chest pain on exertion relieved by rest	___	_____

Chest pain not always associated with exertion	___	_____
Shortness of breath when lying down, relieved by sitting up	___	_____
Unexpected weight loss of more than 10 lbs	___	_____
Unexpected rectal bleeding	___	_____
Leg Pain after walking short distances	___	_____

Women Only (Men May Skip) Please Answer the Following:

	Yes	Date (mo/yr)
Was your last pelvic exam or Pap smear abnormal?	___	_____
Do you have menstrual period problems?	___	
List number of menstrual periods in last year	___	

When was your last menstrual period? (1st day) month _____ day _____ yr _____

Please give lifetime number of: pregnancies _____ living children _____

Men And Women Answer the Following:

	Yes	Date (mo/yr)
Have you ever had:		
A chest x-ray?	___	_____
An abnormal chest x-ray?	___	_____
An ECG (electrocardiogram)?	___	_____
An abnormal ECG?	___	_____
An exercise stress test sometimes called GXT?	___	_____
An abnormal exercise stress test?	___	_____

MEDICATIONS

List medications which you are currently taking on a regular basis, including those prescribed by your doctor and those not prescribed. Give the daily dose and the day/time you last took your medicines. If you don't know the name of a prescribed medicine, please tell what it was prescribed to treat. _____

_____

_____

SURGICAL HISTORY

Check the surgical procedures you have had and give the date of the surgery.

	Yes	Date (mo/yr)
Appendectomy	___	_____
Knee or ankle surgery	___	_____
Arm or shoulder surgery	___	_____
Back surgery	___	_____
Hysterectomy (women only)	___	_____
Vasectomy (men only)	___	_____

Cancer related surgery
  Breast   ___   _____
  Cervix   ___   _____
  Colon   ___   _____
  Lung   ___   _____
  Uterus (women)   ___   _____
  Liver   ___   _____
  Kidney   ___   _____
  Prostate (men)   ___   _____
  Other (Specify) _____   ___   _____

Heart surgery   **Yes**   **Date (mo/yr)**
Heart catheterization   ___   _____
Angioplasty (PTCA)   ___   _____
Coronary bypass (CABG)   ___   _____
Valve repair/replacement   ___   _____
Other _____   ___   _____

ORTHOPEDIC PROBLEMS

Place a check in the blank to indicate any of the following orthopedic problems you may have.

	Yes	Most Recent Occurrence (mo/yr)
Low back pain	___	_____
Shoulder pain	___	_____
Elbow pain	___	_____
Wrist or hand pain	___	_____
Hip problems	___	_____
Knee problems	___	_____
Ankle or foot problems	___	_____
Work or exercise limited by orthopedic problem?	___	_____
Other _____	___	_____

FAMILY HISTORY

Please identify blood relatives who have been diagnosed as having the following diseases and give their age at time of diagnosis.

Heart Disease	Yes	Age at Diagnosis
Father	___	___
Mother	___	___
Sibling	___	___
Paternal grandparent	___	___
Maternal grandparent	___	___

High Blood Pressure	Yes	Age at Diagnosis
Father	___	___
Mother	___	___
Sibling	___	___
Paternal grandparent	___	___
Maternal grandparent	___	___

Stroke	Yes	Age at Diagnosis
Father	___	___
Mother	___	___
Sibling	___	___
Paternal grandparent	___	___
Maternal grandparent	___	___

Have any of your blood relatives noted above had any of the following?

	Yes	Age Diagnosed
Heart attack under age 50	___	___
Heart operations	___	___
Stroke under age 50	___	___
Elevated cholesterol	___	___
High blood pressure under age 40	___	___
Diabetes	___	___
Obesity (BMI ≥ 30)	___	___
Cancer under age 60	___	___

HISTORY OF TOBACCO USE

	Yes	No
Have you ever used tobacco products including smokeless?	___	___
Do you presently use tobacco products?	___	___

If you did or do use tobacco, please indicate the average amount used per day and the age you started.

	Amount	Age Started	Age Quit
Cigarettes (number per day)	___	___	___
Cigars (number per day)	___	___	___
Pipe (number pipefuls per day)	___	___	___
Smokeless (fraction of packs/tins/day)	___	___	___
If you have quit using tobacco, when was it? (mo/yr)	___	___	___
How old were you when you quit using tobacco?	___	___	___

STRESS/TENSION

Rate how closely you agree with each of the following statements by filling in the blank preceding each statement with a number from 1 to 10.

Strongly Disagree　　　**Agree Somewhat**　　　**Strongly Agree**
1　　2　　3　　4　　5　　6　　7　　8　　9　　10

_____ I can't honestly say what I really think or get things off my chest at work, school, or home.
_____ I seem to have lots of responsibilities but little authority.
_____ I seldom receive adequate acknowledgment or appreciation when I do a good job.
_____ I have the impression that I am repeatedly picked on or discriminated against.
_____ I feel I am unable to use my talents effectively or to their full potential.
_____ I tend to argue frequently with coworkers, customers, teachers, or other people.
_____ I don't have enough time for family and social obligation or personal needs.
_____ Most of the time I have little control over my life at work, school, or home.
_____ I rarely have enough time to do a good job or accomplish what I want to.
_____ In general, I'm not particularly proud of or satisfied with what I do.

ALCOHOL CONSUMPTION

Do you drink alcoholic beverages?　　___Yes　　___ No
If **YES,** please estimate the type and amount you consume per week.

	Amount
Glasses of beer per week (12 oz)	_____
Glasses of wine per week (8 oz)	_____
Ounces of liquor (cordials = 1 oz)	_____
Ounces of hard liquor (shot = 1 oz)	_____

SPORT ACTIVITIES

Check those activities in which you regularly participate or in which you have participated over the past year. Also indicate the approximate number of months in the last year you engaged in these activities, the number of times per month, the number of minutes per session, and the intensity of your participation.
Note: Rate your intensity on a scale of **1** to **10** with **1** being very low and **10** being very high intensity.

	No. of months per year	No. times per month	Min/session	Intensity (1 = low; 10 = high)
Basketball	___	___	___	___
Volleyball	___	___	___	___

Softball _____ _____ _____ _____
Baseball _____ _____ _____ _____
Jogging _____ _____ _____ _____
Running _____ _____ _____ _____
Swimming _____ _____ _____ _____
Bicycling _____ _____ _____ _____
Golf _____ _____ _____ _____
Tennis _____ _____ _____ _____
Badminton _____ _____ _____ _____
Racquetball _____ _____ _____ _____
Handball _____ _____ _____ _____
Table Tennis _____ _____ _____ _____
Sailing _____ _____ _____ _____
Water Skiing _____ _____ _____ _____
Horseback Riding _____ _____ _____ _____
Bowling _____ _____ _____ _____
Calisthenics _____ _____ _____ _____
Walking _____ _____ _____ _____
Canoeing/Rowing _____ _____ _____ _____
Fishing _____ _____ _____ _____
Hunting _____ _____ _____ _____
Dancing _____ _____ _____ _____
Skating _____ _____ _____ _____
Soccer _____ _____ _____ _____
Lawnwork/Yard Care _____ _____ _____ _____
Gardening _____ _____ _____ _____
Housework _____ _____ _____ _____
Other _____ _____ _____ _____ _____
Other _____ _____ _____ _____ _____
Other _____ _____ _____ _____ _____

In addition to the above information that you have listed, are you aware of any other conditions, symptoms, or special circumstances that might be related to your overall health and well-being? **Yes or No** __
If **"YES"**, please give a detailed explanation below. _____
_____
_____
_____
_____
_____
_____

THANK YOU VERY MUCH FOR YOUR ACCURANCY IN COMPLETING THIS FORM

——— Informed Consent For Symptom-Limited Maximal Graded Exercise Testing ———
(Modified from Thompson, Gordon, and Pescatello, ACSM's Guidelines for Exercise Testing and Prescription, 2010)

1. **Explanation of Test Procedure**
   You will perform a *maximal* exertion exercise test on a motor-driven treadmill or a stationary cycle ergometer. The exercise intensity will begin at a level you can easily accomplish and will be advanced in stages of increasing difficulty depending on your physical fitness ability. The testing personnel may stop the test at any time because of signs of fatigue or symptoms that may not be considered normal, such as changes in your heartbeat, ECG, or blood pressure. You may choose to stop exercise whenever you wish because of feelings of fatigue or discomfort, or you simply do not wish to continue exercise.

2. **Possible Risks and Discomforts**
   There is the possibility of certain physical changes occurring during the test, some of which may be potentially harmful. These possible changes include abnormal blood pressure, fainting, and disorders of heartbeat. It is very rare, but there is also the possibility of heart attack, stroke, or death during hard exercise. Every effort will be made to minimize the occurrence of these problems by evaluating preliminary information you have provided related to your health and physical fitness, and by careful observations during testing. Emergency equipment is available, and para-medical emergency service is within-miles of this testing facility. Also, trained personnel are available to deal with unusual situations that may arise.

3. **Your Responsibilities**
   To the best of your knowledge you are responsible to truthfully report information about your medical history, current health status, and previous experiences of symptoms which may be related to your heart, or any other symptom that may limit your ability to exercise safely. These include symptoms such as shortness of breath with light physical activity, pain, pressure, tightness, or heaviness in the chest, neck, jaw, back, and/or arms with physical effort. These conditions may affect the safety of your exercise test. It is very important that you promptly report any of these and any other unusual feelings during the exercise test. You should also report to the testing staff all medications prescribed by your doctor and those you take without prescription, as well as the time you last took your medicines.

4. **Benefits You Can Expect**
   The results of this test will be used primarily to provide you with information about your current level of physical working capacity, that is, your physical fitness and cardiorespiratory endurance capacity. The information will also aid in defining an amount of physical activity that is safe and that will help improve your physical fitness. If evaluated by a physician, the results may assist in diagnosing an illness and determining how well your medications are working.

5. **Inquiries about the Test**
   You are encouraged to ask any questions you may have about this maximal exercise test before you begin, and after the test is completed. You should not begin this test until you have your questions answered to your satisfaction.

6. Use of Medical Records

The information obtained during this exercise test will be treated as privileged and confidential in compliance with the Health Insurance Portability and Accountability Act of 1996. No information will be released or revealed to anyone other than your referring physician without written consent from you. You should understand, however, that the information obtained may be used for statistical analysis or scientific purposes with your right of privacy retained.

7. Freedom of Consent for the Test

I voluntarily consent and give my permission to perform this maximal exercise test designed to help me understand more about my physical fitness and cardiovascular health. I understand that I am free to deny consent and to stop the test at any point I so desire.

I have read this consent form, I understand the test procedures that I will perform, and I have been provided an explanation of the possible risks and discomforts. Knowing these risks and discomforts, and having had an opportunity to ask questions that have been answered to my satisfaction, I consent to participate in this submaximal exercise test.

_____ _____
Signature of Participant   Date

_____ _____
Signature of Witness   Date

GRADED EXERCISE TEST WORKSHEET

PRE-EXERCISE DATA		
Subject Name: _____	Date: _____	Test Protocol: _____
Medical History: _____		
Medications/Doses: _____		
Pre-Ex HR: _____ BP: ____/____ Pred Max HR: _____ 85% Pred Max: _____		

EXERCISE DATA

Time	Speed/rpm	Grade/kpm	HR	BP	RPE	$\dot{V}O_2$	Comments

RECOVERY DATA

Time	HR	BP	Comments

MAXIMAL EXERCISE SUMMARY

Max Time: _____   Max HR: _____   Max BP: _____

Pred $\dot{V}O_{2max}$ _____   Measured $\dot{V}O_{2max}$ _____

Stop Reason: _____

1. General/Leg Fatigue
2. Chest Pain/Angina
3. Light headed
4. Dyspnea
5. Claudication
6. Abnormal BP
7. ST $\uparrow \geq 1$ or $\downarrow \geq 2$ mm
8. Induced BB Block
9. V- Tach
10. Sustained SVT
11. 2nd or 3rd Degree HB
12. Frequent PVCs
13. R-on-T PVC
14. Multifocal PVC
15. Other _____

Exercise Technician Name: _____

GRADED EXERCISE TEST REPORT

Demographics
Name: _____ ID #_____ Age: _____ Sex: _____ Date: _____
Height (inches): _____ Weight (pounds): _____ Meds: _____

Pre-Exercise Assessment of Cardiovascular Disease Risk, Health, and Exercise Readiness	
Cardiovascular Disease Risk Factors	**Cardiovascular Disease Status**
_____ Hypertension Blood Pressure _____ mmHg	_____ No known CAD
_____ Total Cholesterol _____ mg/dl	_____ Post M.I.: Date(s) _____
_____ Smoking Pack Years: _____	_____ CABG: Date(s) _____ Vessels: _____
_____ Family History _____	_____ Other Heart Surgery: Date(s) _____
_____ Abnormal Resting ECG _____	Explain: _____
_____ Diabetes	**Symptoms**
_____ Sedentary Lifestyle	_____ Typical Angina _____ Syncope
_____ Obesity (BMI ≥30)	_____ Atypical Angina _____ Dyspnea
_____ Often Feel Emotional Stress	_____ Arrhythmias
_____ Age: Men ≥ 45, Women ≥ 55	_____ Other Chest Pain: Explain: _____
_____ Other _____	Comments: _____
Comments: _____	

Non-Cardiac

Orthopedic Limitations _____   Significant Arthritis _____
Neuromuscular Disorder _____   Acute Illness _____

Resting ECG and Blood Pressure Data	
Supine	HR: _____ BP: _____ ECG*: normal _____ abnormal _____ equivocal _____ inconclusive _____
	Comments: _____

Physical Examination
Cardiovascular

Peripheral Pulses: _____ Normal   _____ Abnormal   Auscultation: _____ Murmurs   _____ Thrill
Evidence of: _____ Edema   _____ Xanthoma   _____ Clicks   _____ Gallop
Bruits: _____ Carotid   _____ Abdominal   _____ Groin   _____ Dysrhythmias
_____ Other: Explain _____   Comments: _____

Pulmonary

_____ Clubbing   _____ Cough   _____ Chest Abnormalities   _____ Wheezes/Rales   _____ Abnormal Spirometry
_____ Other: Explain _____   Comments: _____

GXT Authorization
I have examined the individual name above and **DO or DO NOT** (circle one) approve his/her participation in a symptom-limited maximal graded exercise test and physical fitness evaluation.
Limitations noted: _____
Contraindications or to exercise: _____
Signature or Physician or Test Technologist: _____

(continued)

	Exercise Data	
Pre-Exercise and Maximal Exercise Values	Protocol: _____ PreEx: HR _____ BP ____/____ MHR _____ % Pred Max _____ MBP ____/____ RPE _____ RPP _____ Speed/Grade ____/____ Time _____ $\dot{V}O_{2max}$ _____ METS _____ FAI _____ Reason for Stopping: _____	**Exercise Response Plot**
Exercise Interpretation	*HR responses:* ___ normal ___ abnormal* *BP responses:* ___ normal ___ abnormal* *ECG:* ___ normal ___ abnormal* ___ equivocal* ___ inconclusive* *Symptoms*:*	
Recovery Interpretation	*HR responses:* ___ normal ___ abnormal* *BP responses:* ___ normal ___ abnormal* *ECG:* ___ normal ___ abnormal* ___ equivocal* ___ inconclusive* *Symptoms*:*	

Interpretation Summary and Recommendation

Interpretation/Comments: _____

_____

Recommendation: I have carefully examined this individual's health history, cardiovascular disease risk, and the physiologic and electrocardiographic responses to graded exercise testing and conclude that:

_____ It is SAFE for this individual to participate in exercise prescribed by _____

_____ It MAY NOT BE SAFE for this individual to engage in physical exercise. They should consult personal physician.

Lab Supervisor or Physician: _____ Date: _____

Interpretation Key (*)

ECG		Physiologic Responses	Symptoms
Abnormal	**Equivocal**	Hypotension	Chest pain
ST ↑ ≥ or ↓ ≥ 1 mm	Occasional PVC's	Bradycardia	Dyspnea
Q-waves	First deg. HB	Marked Hypertension	Pallor
Bundle Branch Block	Aberrant A-V conduction	Failure of SBP to rise	Cyanosis
Post-ex. U inv.	Upsloping ST	Other: _____	Lightheadedness
V-Tach	Junctional or atrial arrhythmias	_____	Fainting
Sustained SVT	Other: _____		Other: _____
R-on-T PVC			
Frequent PVC's	**Inconclusive**		
Multifocal PVC's	Submax. effort		
2nd or 3rd deg. HB	Drug effect		
Other: _____	Equipment failure		

(continued)

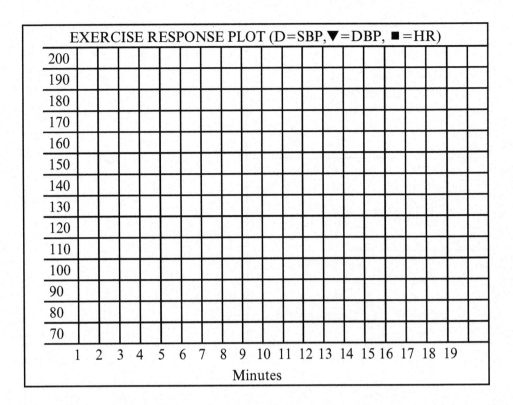

CHAPTER 10

THE MEASUREMENT OF BODY COMPOSITION

OUTCOME OBJECTIVES

Following successful completion of this exercise, the student should be able to:

1. Estimate % body fat using skinfold measurements.
2. Estimate body density and % fat using the underwater weighing technique.
3. Examine the errors inherent in these measurements.

INTRODUCTION

The measurement of body composition is widely performed in exercise physiology and stress testing laboratories. It is one of the primary bases upon which trainers, coaches, and health experts determine the need for weight reduction in their subjects. Body composition estimates are often used for recommending a suitable weight for athletes, setting minimum weight limits for wrestlers, and establishing reasonable weight goals for physical fitness participants. Age-height-weight tables are commonly used for reference standards, but these tables have many inadequacies. They consider only one skeletal dimension (length) and do not take into account variances in body type.

Perhaps the most serious criticism of some age-height-weight tables is that, as one ages, the tables permit increasing fat, without a corresponding increase in body structure (Table 10.1). Humans achieve physical maturation at approximately 25 years of age. Thereafter, unless a person was excessively overweight or underweight at that age, it is desirable to maintain that weight throughout life. The gradual weight gain with advancing years is due to change of eating habits, a more sedentary lifestyle, and a decreasing metabolic rate. Newer tables do not allow this gain, but rather use the same standards for everyone who is of similar build and who is over the age of 25 years (Table 10.2).

Another shortcoming of these tables, especially significant to exercise physiology, is that they make no distinction between fat and lean tissue. Being heavy does not always mean one is fat. Athletes may appear to be heavy when evaluated according to the weight charts, but their "excess" weight may be due to muscle mass, rather than fat. Training tends to increase muscle mass and to reduce fat. Being able to measure body composition and to distinguish between fat and lean tissue is advantageous in such situations.

TABLE 10.1 — Average weights of men and women according to age and height

	A. MEN+							
	Age Groups							
Height	15–16	17–19	20–24	25–29	30–39	40–49	50–59	60–69
5' 0"		98	113	122	128	131	134	136	133
1"	102	116	125	131	134	137	139	136
2"	107	119	128	134	137	140	142	139
3"	112	123	132	138	141	144	145	142
4"	117	127	136	141	145	148	149	146
5"	122	131	139	144	149	152	153	150
6"	127	135	142	148	153	156	157	154
7"	132	139	145	151	157	161	162	159
8"	137	143	149	155	161	165	166	163
9"	142	147	153	159	165	169	170	168
10"	146	151	157	163	170	174	175	173
11"	150	155	161	167	174	178	180	178
6' 0"	154	160	166	172	179	183	185	183
1"	159	164	170	177	183	187	189	188
2"	164	168	174	182	188	192	194	193
3"	169	172	178	186	193	197	199	198
4"	176	181	190	199	203	205	204	

	B. WOMEN+							
	Age Groups							
Height	15–16	17–19	20–24	25–29	30–39	40–49	50–59	60–69
4'10"	97	99	102	107	115	122	125	127
11"	100	102	105	110	117	124	127	129
5' 0"	103	105	108	113	120	127	130	131
1"	107	109	112	116	123	130	133	134
2"	111	113	115	119	126	133	136	137
3"	114	116	118	122	129	136	140	141
4"	117	120	121	125	132	140	144	145
5"	121	124	125	129	135	143	148	149
6"	125	127	129	133	139	147	152	153
7"	128	130	132	136	142	151	156	157
8"	132	134	136	140	146	155	160	161
9"	136	138	140	144	150	159	164	165
10"		142	144	148	154	164	169	
11"			147	149	153	159	169	174
6' 0"			152	154	158	164	174	180

From L. J. Bogert, G. M. Briggs, and D. H. Calloway. *Nutrition and Physical Fitness,* 9th ed., Philadelphia: W.B. Saunders, 1973, excerpted from *Build and Blood Pressure Study,* Society of Actuaries, 1959.
+Graduated weights in indoor clothing in pounds.

TABLE 10.2 — Desirable weight for men and women over 25 years of age

A. MEN +				
Height (with shoes on) 1-inch heels				
Feet	Inches	Small Frame	Medium Frame	Large Frame
5	2	112–120	118–129	126–141
5	3	115–123	121–133	129–144
5	4	118–126	124–136	132–148
5	5	121–129	127–139	135–152
5	6	124–133	130–143	138–156
5	7	128–137	134–147	142–161
5	8	132–141	138–152	147–166
5	9	136–145	142–156	151–170
5	10	140–150	146–160	155–174
5	11	144–154	150–165	159–179
6	0	148–158	154–170	164–184
6	1	152–162	158–175	168–189
6	2	156–167	162–180	173–194
6	3	160–171	167–185	178–199
6	4	164–175	172–190	182–204

B. WOMEN ++				
Height (with shoes on) 2-inch heels				
Feet	Inches	Small Frame	Medium Frame	Large Frame
4	10	92–98	96–107	104–119
4	11	94–101	98–110	106–122
5	0	96–104	101–113	109–125
5	1	99–107	104–116	112–128
5	2	102–110	107–119	115–131
5	3	105–113	110–122	118–134
5	4	108–116	113–126	121–138
5	5	111–119	116–130	125–142
5	6	114–123	120–135	129–146
5	7	118–127	124–139	133–150
5	8	122–131	128–143	137–154
5	9	126–135	132–147	141–158
5	10	130–140	136–151	145–163
5	11	134–144	140–155	149–168
6	0	138–148	144–159	153–173

+Weight in pounds according to frame (in indoor clothing).
++For girls between 18 and 25, subtract 1 pound for each year under 25.

When body composition is measured the body is divided into two components, fat-free mass and stored fat. The student should become familiar with the following terminology:

<u>Fat free mass</u> (FFM): All of the body tissue with the exception of stored ("depot") fat. It includes muscle, bone, and nerve fiber coverings, as well as the essential fats the body must have for cell wall construction, and other structures.

<u>Fat free weight</u> (FFW): The quantitative expression of the FFM. FFW is usually expressed in kilograms or pounds. FFW is equal to the total body weight minus the weight of the stored fat.

Most changes in body weight in adults are due to changes in fat content, while the FFW remains relatively constant or decreases slightly with age. Fat content varies in different groups. College-age men average about 15% of the total weight as fat, while college-aged women average about 23%. Men and women training for many sports will tend to have less fat than their untrained counterparts. Highly trained endurance athletes are the best example, sometimes having as little as 6 to 8% fat. It is absolutely normal to have some fat as a nutritional reserve, however, and levels less than this are generally considered unhealthy (Friedl et al., 1994).

There are a number of methods for measuring body fat percentage. Most are based on very sound physiological principles and have the potential to provide reasonable results. Recently, several scanning techniques, most commonly Dual-energy X-ray Absorptiometry (DEXA or DXA), have come into more widespread usage with the decreased cost of the equipment. Evidence indicates that these may be very accurate (Friedl et al., 1994). They appear to provide higher body fat estimates than other methods, so some evaluation may still be required (Borrud et al., 2010; Li et al., 2009). Still, even the best of techniques in the most expert hands, is *only an estimate,* not a true measurement. This exercise is designed to illustrate two of the primary techniques, hydrostatic weighing and skinfold thickness measurement, used in body composition analysis and to allow you to practice performing the calculations used in the measurements.

Equipment Needed

Skinfold caliper

Water tank or swimming pool

Scale for subject weight

Hanging metric scale (24 kg × 50 grams)

TECHNIQUES AND PROCEDURES

SKINFOLDS

Many techniques for measuring body fatness are frequently inappropriate for use in physical education classes or coaching, due to cost or technical considerations. Consequently, simplified methods using skinfold thicknesses are often employed. Skinfold assessments of body fat are based on the assumption that measurements of subcutaneous fat at various body sites can be used to estimate total body fat. In general, skinfold measurements are taken at specific sites on the surface of the body, summed, then entered into specific equations to predict body density. Sites are chosen to reflect places on the body where we tend to deposit subcutaneous fat, as well as sites where we do not, to get an accurate reflection of subcutaneous fat over the body. Once an estimated body density is known, body fat percentage can be calculated.

Although this method has an advantage of simplicity over more involved techniques and is less expensive to perform, it is subject to sizable errors if performed incorrectly. These errors can be brought about by the assumption of fat deposition patterns, the density of lean body mass, and the tension on the spring calipers, among others. Can you think of other potential errors that might be associated with measuring skinfolds? Even in the most expert hands, the measurement error has been estimated to be about 3.7% (Pollock and Wilmore, 1990, p. 334). There are well over 100 population-specific equations to estimate body fat from skinfolds, circumference measured, diameters of bony structures, or some combination of these variables. The procedures for using two generalized equations for men and women, one requiring measurement of skinfold thicknesses at seven sites and one at three sites are detailed below. The following procedure is recommended:

1. With a felt-tip pen, carefully mark the anatomical location of the skinfold measures to be taken (see below.) All measures should be taken on the right side of the body. If the subject has had an injury resulting in scarring in places where the skinfolds would be taken on the right side, this could interfere with the measurement. In such a case it is acceptable to use the left side. This should be noted, though.
2. Pick up the subject's skinfold between your index finger and thumb of your left hand. Be sure that you have two layers of skin and the underlying fat *only*, and not muscle. Allow the skinfold to follow the natural stress lines of the body. If you doubt that you have a correct skinfold, have the subject contract the underlying muscle; if you have a correct skinfold you will be able to retain your grasp. The fold should be held between the fingers when the measurement is made.
3. Apply the jaws of the caliper perpendicular to the fold about 1 cm from the fingers. Release the jaw pressure slowly. The pressure on the fold must be exerted by the calipers only!
4. Read the measurement within 3–4 sec of applying the caliper pressure. Measure all skinfolds to the nearest 0.5 mm. Take the *mean* of two measures which are within 5% accuracy (± 1 mm). Take all measures once and then repeat the entire sequence until two measures of the desired accuracy are obtained.
5. Record your data *accurately* on your worksheet.

ANATOMICAL LANDMARKS FOR SKINFOLD MEASURES

a. Chest—diagonal fold along the lateral border of the pectoralis muscle one-third (women) or one-half (men) of the distance between the anterior axillary line and the nipple. 3, 3

b. Midaxilla—vertical fold along the midaxillary line at the level of the xiphoid process (lower border) of the sternum. 5, 4.5

c. Subscapular—diagonal fold 1–2 cm from the inferior angle of the scapula. 7.5, 7.5

d. Triceps—vertical fold halfway between the acromion and olecranon processes on the posterior midline of the upper arm. The arm should be relaxed and extended at the side. 4, 4.5

e. Abdomen—vertical fold level with the umbilicus, about 2 cm to one side. 8.5, 7

f. Suprailium—diagonal fold at the iliac crest and the anterior axillary line. 5, 5

g. Thigh—vertical fold on the anterior surface of the thigh, half way between the proximal border of the patella and the intersection of the inguinal crease and the midpoint of the long axis of the thigh. (See Figure 10.1.) 5, 5

FIGURE 10.1 — Sites used in performing skinfold measurement of body composition. A – Triceps; B – Chest; C – Midaxillary; D – Subscapular; E – Suprailiac; F – Abdomen; G – Thigh

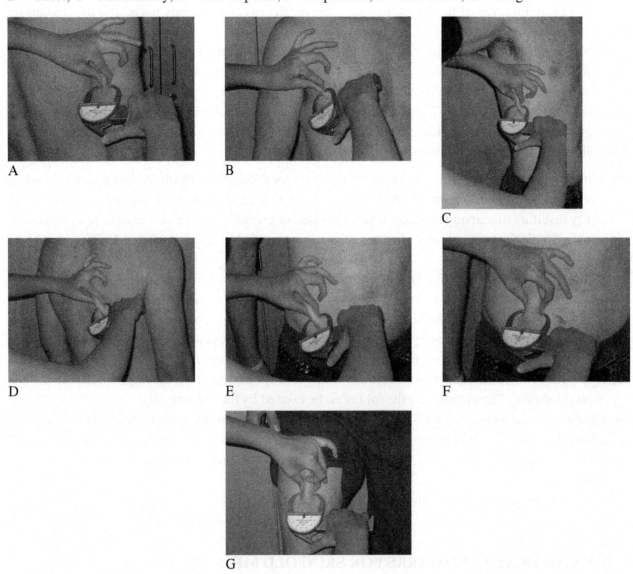

6. From these measurements, you can now compute the body density using one of the following formulae:

Females

$$D_b = 1.0970 - 0.00046971(\Sigma 7) + 0.00000056(\Sigma 7)^2 - 0.00012828(A)$$

$$D_b = 1.0994921 - 0.0009929(\Sigma 3) + 0.0000023(\Sigma 3)^2 - 0.0001392(A)$$

Males

$$D_b = 1.1120 - 0.00043499(\Sigma 7) + 0.00000055(\Sigma 7)^2 - 0.00028826(A)$$

$$D_b = 1.10938 - 0.0008267(\Sigma 3) + 0.0000016(\Sigma 3)^2 - 0.0002574(A)$$

Where: D_b = Body Density; $\Sigma 7$ = Sum of seven skinfolds (mm); $\Sigma 3$ (female) = Sum of three skinfolds (triceps, suprailiac, and thigh); $\Sigma 3$ (male) = Sum of three skinfolds (chest, abdomen, and thigh); A = Age (years). (Pollock, Schmidt, and Jackson, 1980)

UNDERWATER WEIGHING

In this method, sometimes called hydrostatic weighing, body density (D_b) is measured directly. D_b can be calculated as body weight divided by body volume. It is relatively simple to measure body weight with reasonable accuracy using a laboratory scale. Measuring body volume (V_b) is more difficult. In the underwater weighing technique, Archimede's principle is applied to this problem. This principle states that any body immersed in water is acted upon by a buoyant force which is equal to the weight of the displaced water. This buoyant force is evident as a loss of weight in water. Therefore, the weight of water displaced by a submerged body can be determined by computing the difference between the weight of the body in the air and the weight of the body in the water. Water density varies with water temperature (1 gm ml^{-1} at 4°C), but can be readily obtained from published tables. Once the water density and the weight of the water displaced is known, the volume of the water displaced can be calculated; V_b is assumed to be equal to the volume of the displaced water. Residual volume of the lungs and the volume of gas in the GI tract are subtracted from Vb, since these volumes do not contribute to the body weight.

The density of body fat is less than water (0.9), whereas that of bone and muscle is higher than water (1.2 to 3.0). Therefore, as the proportion of body fat increases, D_b is reduced. In simple terms, people with relatively higher amounts of body fat tend to float, while those with proportionally more muscle tissue tend to sink in water. Although this technique is one of the most widely used for measuring body density, it is important to note that the standard error of this procedure has been estimated to be about 2.7% (Lohman, 1981). Along with technician error, the use of predicted rather than measured residual volume can detract substantially from the measurement accuracy of this technique. A step-by-step procedure for underwater weighing is given below.

1. Upon scheduling a subject for this procedure, instruct them to abstain from eating for at least 2–3 hours prior to the test.
2. The tank should be clean and filled with fresh, warm (32–36° C) water.
3. When the subject arrives, carefully explain all procedures and have them sign an informed consent.
4. Determine the residual volume. For this lab, use the gender-specific equation by Goldman which is given on the worksheet under the section titled, "Predicted Residual Volume." Residual volume can also be measured using the technique described in the Pulmonary Function chapter.
5. Weigh the bathing suit in the air on the lab scale. If the weight is negligible, disregard it in the calculations.
6. Have the subject retire to the dressing room, change into bathing suit, void the bladder and bowels, and report back to the testing area.
7. Carefully record the height and weight of the subject wearing the bathing suit only. Record the weight to the nearest 1/4 pound.
8. Instruct the subject to retire to the dressing room and shower thoroughly then report back to the testing area. This removes dead skin cells, body oil, and loose hair, preventing them from collecting in the weighing tank.
9. Have the subject enter the tank and sit comfortably in the chair. The water level should be about shoulder depth when seated.
10. Measure and record the water temperature just prior to each test, even if several are done in sequence. For subject comfort, the water temperature should be between 32° and 36°C. Note and record the water density at the measured temperature (see chart below) on your worksheet.

Water Temp (°C)	Water Density
30	0.995678
31	0.995372
32	0.995057
33	0.994734
34	0.994734
35	0.994063
36	0.993716
37	0.993360

11. Instruct the subject carefully in all procedures before beginning the actual measurement.
12. Instructions to subject:
    a. Submerge and shake your hair free of all bubbles. Press all of the air bubbles out of your bathing suit as well.
    b. Attach nose clip if desired and hyperventilate for 4 to 5 breaths, then exhale most of the air with your head still out of the water.
    c. Submerge *slowly* and completely while expiring remaining air from your lungs. While totally submerged, completely force *all* the air out of your lungs. The technician will record your underwater weight when satisfied that you have exhaled as much air as possible (no more air bubbles appear around your mouth). You will repeat this procedure a total of seven to ten times.
13. Instruction to technician:
    a. Instruct the subject to submerge and surface *slowly* so as to minimize wave formation and expedite the procedure.
    b. Be sure the subject is totally submerged, expels all air completely, and is not touching the sides or bottom of the tank.
    c. Quickly and carefully record the underwater weight of the subject from the scale. When done, shake the scale to cue the subject to surface.
    d. Take a minimum of seven weights and record the mean of the highest three weights (within ± 1%) as the actual underwater weight. When successive weights plateau, even with encouragement from the tester, the test can be terminated.
    e. <u>Remember</u>: Record the tare weight (weight of the chair including the bathing suit) after the test.
    f. Calculate body density (D_b) using the following formula:

    $$D_b = \frac{Wta}{\frac{Wta - Wtw}{Dw} - (RV + 0.1L)}$$

    Where: Wta = body weight in air
    Wtw = body weight in water
    Dw = water density
    RV = residual volume
    0.1L is the assumed volume of gas in the GI tract

    g. Record the data on your data sheet. (See Figure 10.2.)

FIGURE 10.2 — Photograph of underwater weighing apparatus.

BODY FAT

The previously described techniques were used to estimate total body density. Body fatness can be estimated from body density based on the principle that fat is less dense that lean weight (0.9 vs 1.1 gm/ml). One of the following estimates can be used to convert body density to percent body fat:

$$\% \text{ Fat} = (4.950/D_b - 4.500) \times 100 \text{ (Siri, 1961)}$$

$$\% \text{ Fat} = (4.570/D_b - 4.142) \times 100 \text{ (Brozek et al., 1963)}$$

Because of differences in the masses of skeletal muscle and bone minerals, some races have different body densities with the same percentage of fat. Therefore, equations have been developed for people of African descent to account for this difference:

$$\% \text{ Fat} = 4.832/D_b - 4.369 \times 100 \text{ (Black females. Ortiz et al., 1992)}$$

$$\% \text{ Fat} = 4.374/D_b - 3.928 \times 100 \text{ (Black males. Schutte et al., 1984)}$$

Calculate percent body fat on yourself using both the skinfolds and the underwater weighing values given. Also calculate fat-free weight and target body weight.

QUESTIONS AND ACTIVITIES

1. Make a table of class results: gender, percent body fat skinfold, percent body fat underwater weighing.
2. Graph the average results for both techniques using gender as the independent variable.
3. What is the average difference between the two estimates for each gender?
4. Are the skinfold estimates always higher or lower than those obtained from underwater weighing? Why or why not?
5. List reasons for errors that could be made in each method.

REFERENCES

Behnke, A. R. "Physique and Exercise," Chapter 12 in Falls, H. B. (Ed.). *Exercise Physiology,* New York: Academic Press, 1968, pp. 359–86.

Brozek, J., F. Grande, T. Anderson, and A. Keys. "Densitometric Analysis of Body Composition: Revision of Some Quantitative Assumptions." *Ann N Y Acad Sci.* 110 (1963): 113–40.

Borrud, L. G., K. M. Flegal, A. C. Looker, J. E. Everhart, et al. "Body Composition Data for Individuals 8 Years of Age and Older: U.S. Population, 1999–2004." National Center for Health Statistics. *Vital Health Stat* 11 (2010): 1–87.

deVries, H. A. *Physiology of Exercise for Physical Education and Athletics,* 2nd ed., Dubuque, IA: Wm C. Brown, 1974, pp. 250–55.

Friedl, K. E., R. J. Moore, L. E. Martinez-Lopez, J. A. Vogel, E. W. Askew, L. J. Marchitelli, R. W. Hoyt, and C. G. Gordon. "Lower Limit of Body Fat in Healthy Active Men." *J. Appl. Physiol.* 77 (1994): 933–40.

Ismail, A. H., "Body Composition and Relationships to Physical Activity," Chapter 13 in Falls, H. B. (Ed.), *Exercise Physiology,* New York: Academic Press, 1968, pp. 387–92.

Karpovich, P. V. and W. E. Sinning. *Physiology of Muscular Activity,* 7th ed., Philadelphia: W.B. Saunders, 1971, pp. 295–313.

Li, C., E. S. Ford, G. Zhao, L. S. Balluz and W. H. Giles. "Estimates of Body Composition with Dual-Energy X-Ray Absorptiometry in Adults." *Am J Clin Nutr.* 90 (2009): 1457–65.

Lohman, T. G. "Skinfolds and Body Density and Their Relation to Body Fatness: A Review." *Human Biology.* 53 (1981): 181–225.

Ortiz, O., M. Russell, T. L. Daley, R. N. Baumgartner, M. Waki, S. Lichtman, J. Wang, R. N. Pierson, Jr., and S. B. Heymsfield. "Differences in Skeletal Muscle and Bone Mineral Mass Between Black and White Females and Their Relevance to Estimates of Body Composition." *Am J Clin Nutr.* 55 (1992): 8–13.

Pollock, M. L., D. H. Schmidt, and A. S. Jackson. "Measurement of Cardiorespiratory Fitness and Body Composition in the Clinical Setting." *Comprehensive Therapy.* 6.9 (1980): 12–27.

Pollock, M. L. and J. H. Wilmore. *Exercise in Health and Disease,* 2nd ed. Philadelphia: W.B. Saunders, 1990.

Ricci, B. *Physiological Basis of Human Performance.* Philadelphia: Lea & Febiger, 1967, pp. 270–73.

Schutte J. E., E. J. Townsend, J. Hugg, R. F. Shoup, R. M. Malina, and C. G. Blomqvist. "Density of Lean Body Mass Is Greater in Blacks Than in Whites." *J Appl Physiol.* 56 (1984): 1647–49.

Siri, W. E. "Body Composition from Fluid Spaces and Density." In Brozek, J. and A. Henschel (eds.). *Techniques for Measuring Body Composition.* Washington, D.C., National Academy of Science, 1961, pp. 223–44.

INDIVIDUAL WORKSHEET FOR BODY COMPOSITION

Tech name _____   Date _____

Subject name __Jason Farron__   Age __25__

Height __183__ (cm)   Weight[1] in Air (WT_a) __97.9__ (kg)

Skinfold Assessment of Body Fat

A. Skinfold data

Trial (mm)	Chest	Midaxillary	Triceps	Subscapular	Abdominal	Suprailiac	Thigh
1	19	21	15	19	21	20	23.5
2	21	23	16	21	21	22	25
Average[2]	20	22	15.5	20	21	21	24.25

143.75

B. Calculation of Body Density (D_b) from Skinfolds

1. Compute the sum of skinfolds using average values

Adult Men (18–61 years)								
Fold	Chest	Midaxillary	Triceps	Subscapular	Abdominal	Suprailiac	Thigh	Sum
Σ7								143.75
Σ3	20	X X X	X X X	X X X	21	X X X	24.25	65.25

Adult Women (18–55 years)								
Fold	Chest	Midaxillary	Triceps	Subscapular	Abdominal	Suprailiac	Thigh	Sum
Σ7								
Σ3	X X X	X X X		X X X	X X X			

2. Using the appropriate skinfold sums, apply the gender-specific formula below to calculate body density.

Adult Men

$\Sigma 7\ D_b = 1.11200000 - 0.00043499\ (\Sigma 7\_\_) + 0.00000055\ (\Sigma 7\_\_)^2 - 0.00028826\ (Age) = \_\_\_$ kg/L

$\Sigma 3\ D_b = 1.1093800 - 0.0008267\ (\Sigma 3\_\_) + 0.0000016\ (\Sigma 3\_\_)^2 - 0.0002574\ (Age) = \_\_\_$ kg/L

Adult Women

$\Sigma 7\ D_b = 1.09700000 - 0.00046971\ (\Sigma 7\_\_) + 0.00000056\ (\Sigma 7\_\_)^2 - 0.00012828\ (Age) = \_\_\_$ kg/L

$\Sigma 3\ D_b = 1.0994921 - 0.0009929\ (\Sigma 3\_\_) + 0.0000023\ (\Sigma 3\_\_)^2 - 0.0001392\ (Age) = \_\_\_$ kg/L

C. Calculation of Body Fat Percentage (% Fat), Fat Weight (FW), Fat-Free Weight (FFW), and Target Weight (TW)

1. Calculate % Fat

    $\Sigma 7\ \%\ Fat = [(4.570/D_b\_\_\_\_\_) - 4.142] \times 100 = _____$ % Fat

    $\Sigma 3\ \%\ Fat = [(4.570/D_b\_\_\_\_\_) - 4.142] \times 100 = _____$ % Fat

2. Compute FW

    $\Sigma 7\ FW = WT_a\_\_\_\_\_ \times (\%\ Fat\_\_/100) = _____$ kg

    $\Sigma 3\ FW = WT_a\_\_\_\_\_ \times (\%\ Fat\_\_/100) = _____$ kg

3. Compute FFW

    $\Sigma 7\ FFW = WT_a\_\_\_\_\_ - FW\_\_\_\_\_ = _____$ kg

    $\Sigma 3\ FFW = WT_a\_\_\_\_\_ - FW\_\_\_\_\_ = _____$ kg

4. Calculate TW. (Desired % Fat 15% for men and 22% for women.)

    ### Men

    $\Sigma 7\ TW = FFW\_\_\_\_\_ /.85 = _____$ kg

    $\Sigma 3\ TW = FFW\_\_\_\_\_ /.85 = _____$ kg

    ### Women

    $\Sigma 7\ TW = FFW\_\_\_\_\_ /.78 = _____$ kg

    $\Sigma 3\ TW = FFW\_\_\_\_\_ /.78 = _____$ kg

5. Calculate Recommended Weight Loss[3] (RWL)

    $\Sigma 7\ RWL = WT_a\_\_\_\_\_ - TW\_\_\_\_\_ = _____$ kg

    $\Sigma 3\ RWL = WT_a\_\_\_\_\_ - TW\_\_\_\_\_ = _____$ kg

UNDERWATER WEIGHING

3.5 kg

A. Calculate Predicted Residual Volume (RV)

Men: $RV = (0.0272 \times Height\_\_\_\_cm) + (0.017 \times Age\_\_\_\_yrs) - 3.48 = \_\_\_\_L$

Women: $RV = (0.032 \times Height\_\_\_\_cm) + (0.009 \times Age\_\_\_\_yrs) - 3.90 = \_\_\_\_L$

B. Measure Body Volume

1. Underwater Body + Chair Weight (UB)

Trial	1	2	3	4	5	6	7
UB							

2. Average of Highest Three Trials[4]: $UB_a = \underline{3.5}$ kg
3. Tare Weight (Chair + Swimming Suit): $WT_T = \underline{1.15}$ kg
4. Underwater Body Weight (WT_w) =
   $UB_a \underline{3.5} - WT_T \underline{1.15} = \underline{2.4}$ kg
5. Water Temperature = $\underline{32}$ C°
   Water Density (D_w) = ____ kg L^{-1} (or gm ml^{-1}) $\underline{0.995057}$
6. Calculate Body Volume (V_b). (Note: Correction for GI Tract gas = 0.1 L)
   $V_b = [(WT_a \underline{97.9} - WT_w \underline{2.4})/D_w \underline{}] - (RV\_\_\_\_ + 0.1\ L) = \_\_\_\_L$
7. Compute Body Density $\underline{0.995057}$ $\underline{1.9226}$
   $D_b = WT_a \underline{97.9} / V_b \underline{93.95} = \_\_\_\_$ kg L^{-1}

C. Calculate %Fat, FW, FFW, TW, and RWL

1. %Fat = $[(4.570 \div D_b\_\_\_\_) - 4.142] \times 100 = \_\_\_\_\%$
2. FW = $WT_a\_\_\_\_ \times (\%Fat\_\_\_\_\%/100) = \_\_\_\_$ kg
3. FFW = $WT_a\_\_\_\_ - FW\_\_\_\_ = \_\_\_\_$ kg
4. Target Weight (TW)
   Men: TW = FFW____/0.85 = ____ kg (15% Fat)
   Women: TW = FFW____/0.78 = ____ kg (22% Fat)
5. RWL[3] = $WT_a\_\_\_\_ - TW\_\_\_\_ = \_\_\_\_$ kg
   or = (kg____ × 2.2 = ____ lbs.)

---

[1] Unless negligible, bathing suit weight should be subtracted from WT_a.
[2] Average of two measures that meet the accuracy of ±5%.
[3] Applies ONLY when TW is less than WT_a.
[4] Three weights should be within ±1%.

CHAPTER 11

EXERCISE PRESCRIPTION FOR AEROBIC FITNESS IN HEALTHY POPULATIONS

OUTCOME OBJECTIVES

Following successful completion of this exercise, the student should be able to:

1. Assess and evaluate individual needs, goals, health status, and physical activity history prior to prescribing exercise for apparently healthy individuals.
2. Design an individualized exercise prescription program to improve or maintain cardiorespiratory fitness using the five component model of exercise prescription, which consists of mode, frequency, intensity, duration, and progression of exercise.
3. Make recommendations regarding the proper organization of an exercise session with consideration given to the warm-up, the conditioning phase, and the cool-down.

INTRODUCTION

As scientific evidence accumulates, it is becoming increasingly clear that regular physical exercise contributes to optimal health. The evidence is so overwhelming with respect to coronary artery disease (CAD) that the American Heart Association recognizes physical inactivity as a primary cardiovascular disease risk factor along with smoking, high blood pressure, and high blood cholesterol (Fletcher et al., 1992). Physical activity also has beneficial effects on other recognized CAD risk factors. Some studies have shown that exercise can raise high-density lipoprotein-cholesterol and lower triglycerides in the blood (Crouse et al., 1997; Thompson et al., 2001), lower blood pressure (Hagberg, Park, and Brown, 2000), promote fat loss while sparing lean body mass (Weinheimer, Sands, and Campbell, 2010), and can lower the risk for the development of diabetes (Sanz, Gautier, and Hanaire, 2010). Furthermore, individuals who are physically fit also have a reduced risk of developing some types of cancer (Rogers et al., 2008). Not only does regular exercise decrease the risk of some diseases, it also contributes to the overall functional health of many bodily systems. Several of the beneficial effects of exercise are shown in Table 11.1.

Despite the evidence linking physical activity to improved health, surveys have shown that about one in four Americans aged 18 and older do not engage in any form of leisure-time physical activity (Centers for Disease Control and Prevention, 2005). Furthermore, less than 50% of American adults meet the physical activity recommendations published jointly by the U.S. Center for Disease Control and the American College of Sports Medicine (Pate et al., 1995; Thompson, Gordon, and Pescatello, 2010). Although there has been a trend since 1994 for an increased involvement by Americans in leisure-time physical activities, it is apparent that significant strides in improving public health could be made by encouraging even more adults to become physically active (Centers for Disease Control and Prevention, 2005).

TABLE 11.1 — Beneficial physiologic changes with training

System	Variable	Change with Training
Skeletal Muscle	Myoglobin stores	Increase
	Glycogen and triglyceride storage	Increase
	Number and size of mitochondria	Increase
	Oxidative enzyme activities	Increase
	Fiber size and muscular strength	Increase
Cardiorespiratory System	Maximal oxygen uptake	Increase
	Resting heart rate	Decrease
	Heart size and volume	Increase
	Blood volume and total hemoglobin	Increase
	Submaximal exercise heart rate	Decrease
	Stroke volume-submaximal and max	Increase
	Maximum cardiac output	Increase
	Submaximal systolic blood pressure	Decrease
	Resting blood pressure	Decrease
	Arteriovenous oxygen difference	Increase
Other	High-density lipoprotein	Increase
	Low-density lipoprotein	Decrease
	Triglycerides	Decrease
	Bone, tendon, ligament strength	Increase
	Body fat	Decrease

Beneficial Changes in Physiologic Factors Resulting from Physical Training.

Planned and regular physical activity (exercise) scientifically applied in an appropriate manner can beneficially impact cardiovascular and respiratory fitness, muscular strength and endurance, joint flexibility, and body composition—all components of health-related physical fitness. The prioritized and specific component targeted for training should be considered in the prescription process, since specificity of training requires that each receives attention to optimize the training outcome. One component may be emphasized over another depending on subject goals, previous training experience, age, physical ability, and the like. In this laboratory chapter, the focus will be on a predominately aerobic exercise prescription, since this is typically the priority for health-related benefits in the general population.

The amount and type of exercise necessary to enhance health-related fitness can be thought of in terms of a daily prescription, an exercise prescription, analogous to a recommended daily allowance or dose of vitamins. Five components comprise the building blocks of the exercise prescription process. These components can be applied in developing an exercise prescription no matter what the person's age, disease status, or current fitness level. The general recommendations for each component of the exercise prescription paradigm to improve aerobic fitness are as follows (Thompson, Gordon, and Pescatello, 2010):

1. **Mode** — the type or kind of exercise. For aerobic training, include any activity that is rhythmic, uses large muscle groups, can be maintained for prolonged periods of time, and is aerobic in nature.

Examples include walking, jogging, running, swimming, skating, bicycling, rowing, cross-country skiing, hiking, dancing, stair-climbing, and various endurance sport or game activities.

2. **Frequency** — the number of exercise sessions each day or week. Health benefits can accrue when inactive individuals adopt a more active lifestyle, accumulating some exercise most days of the week, even if the physical activities are relatively easy to perform. Added benefits are realized when individuals engage in physical efforts at a higher intensity at least 3 to 5 d · wk^{-1}. This is generally the goal of most exercise prescription programs.

3. **Intensity** — the physical difficulty of the exercise or the physical effort required. There is general agreement that exercise must be at least of moderate intensity to optimize health-related benefits. Intensity is often given as at a percentage of maximal aerobic capacity. In practice, an effort range from 40% to 85% of VO$_{2max}$, from 40% to 85% of maximal heart rate reserve, or from 55% to 90% of maximal heart rate is usually recommended. Lower intensities may suffice for improving health and increasing fitness in those who are sedentary or poorly fit, but will not result in significant cardiorespiratory training benefits over time. Those who are more physically fit will require exercise at the higher end of the range to improve.

4. **Duration or Time** — the length of time the exercise should be continued each session or accumulated over a week. About 20 to 30 minutes of moderate to vigorous exercise performed 3 to 5 d · wk^{-1}, totaling to 75 to 150 min · wk^{-1}. If weight loss is a goal of exercise, 50 to 60 min · d^{-1} at moderate to vigorous exercise is required to total 150 min · wk^{-1} (vigorous) to 300 min · wk^{-1} (moderate) exercise. Equal benefits are realized when the total exercise time is reached by performing intermittent exercise of at least 10 minutes in duration over the course of the day.

5. **Progression** — the rate at which the exercise dose is increased. Systematically increase the dose of exercise per session and per week to accommodate improvements in physical fitness and maintain the overload to stimulate further improvement. This is accomplished by increasing the exercise intensity, duration, or frequency, that is, volume of exercise, in a manner consistent with the safety and goals of the individual. The prescription for rate of progression can be divided into three stages: (1) initial conditioning stage, (2) improvement conditioning stage, and (3) maintenance conditioning stage.

The dose of exercise, sometimes also termed "volume of exercise," can be quantified by expressing the energy expenditure of each exercise session in kilocalories (kcal or Calorie). It should be evident that the frequency, intensity, and duration components can be manipulated to achieve individualized doses of exercise. In other words, energy expenditure varies directly with these three components of the exercise prescription. The dose of exercise seems to be the critical factor responsible for a change in fitness (Blair et al., 1992; Fletcher et al., 1992), even when exercise is performed intermittently throughout the day. For example, there was no difference in fitness changes after eight weeks of training between individuals who completed a single 30-minute exercise session once per day and those who completed three 10-minute sessions per day at an equivalent intensity (Debusk et al., 1990). Also, those who train at different intensities, yet maintain equivalent caloric expenditures show similar improvements in fitness and in other health-related factors, such as blood lipids (Crouse et al., 1997; Wenger and Bell, 1986). Evidence has accumulated over the years to show there is a dose-response relationship between caloric expenditure in physical activity and benefits to health and fitness. The minimum dose threshold for optimizing health/fitness benefits appears to average about 1,000 kcal of physical activity/exercise per week, and this has come to be the recommended minimum quantity of physical activity for healthy adults. For the average adult, this equates to about 150 min · wk^{-1} or about 200 to 350 kcal · day^{-1} given a 3 to 5 d · wk^{-1} frequency (Haskell et al., 2007; Lee, 2010; Thompson, Gordon, and Pescatello, 2010).

When a previously sedentary individual first begins to exercise, lower energy expenditure per session will likely prove to be a sufficient stimulus to promote improvements in fitness.

It should be recognized that vigorous exercise is not without risk, even for the apparently healthy, although the actual risk is quite low. When those with known heart disease are excluded, death during vigorous exercise has been estimated at 1 death per year for every 15,000 to 18,000 exercising adult men. In cardiac patients exercising in cardiac rehabilitation programs 1 nonfatal and 1 fatal cardiovascular complication occurs per 34,673 and 116,402 patient-hours, respectively. The death rate is much lower in the youth estimated to be 1 death per 133,000 high school/college male athletes and even less, 1 in 769,000 women athletes (Thompson et al., 2007). However, this transient increase in risk of death during exercise must be balanced against the facts pointed out previously that physically active individuals are, overall, *less* likely to suffer death from heart disease and cancer. When considered in this light, the risk/benefit ratio for physical activity is highly favorable, and sedentary individuals should be encouraged to become active.

Since some risk is associated with exercise, it is prudent to obtain sufficient information to determine an individual's present health status, risk of disease, physical fitness, and special needs, prior to making an exercise prescription. This type of information will serve as a basis for making decisions regarding: (1) the safety of exercise for the individual and possible exclusion from participation in a non-medically supervised exercise program; (2) the need for a comprehensive medical evaluation and exercise test before starting an exercise program; (3) the beginning dose of exercise which is appropriate; and (4) special considerations to take into account when prescribing exercise to the individual. The American College of Sports Medicine has defined three health classification categories, based on an assessment of health, risk factor, and fitness status, for those who plan to increase their physical activity (Thompson, Gordon, and Pescatello, 2010). These risk categories are:

1. **Low risk** — men and women who have no known disease or disease symptoms, and who have no more than one cardiovascular disease risk factor (Table 11.2).
2. **Moderate risk** — men and women without known disease or disease symptoms, but who have two or more cardiovascular disease risk factors (Table 11.2).
3. **High risk** — those with known cardiovascular, pulmonary, or metabolic disease *or* who exhibit symptoms and signs of these diseases.

Methods of obtaining health and disease risk information and written forms/questionnaires to acquire this information have been carefully described in Laboratory 8, "Submaximal Exercise Testing for Estimating Aerobic Capacity" and Laboratory 9, "The Symptom-Limited Maximal Graded Exercise Test for Clinical and Sports Medicine Applications." In brief, the assessment information can be obtained from an individual in several ways: (1) through a one-on-one interview, (2) by using a detailed health history questionnaire such as the Health and Lifestyle History form which accompanies Laboratory 9, "Symptom-Limited Maximal Graded Exercise Test for Clinical and Sports Medicine Applications," or (3) by using a simpler questionnaire, such as the one included in this laboratory entitled Exercise Prescription Subject Questionnaire.

Once the risk category has been determined for an individual, a decision can be made relative to a requirement for a medical examination and diagnostic or functional symptom-limited maximal graded exercise test prior to beginning exercise or developing an exercise prescription. As shown in Table 11.3, this decision will be based to some extent on whether the individual plans to begin a moderate or vigorous exercise program. Moderate exercise has been defined as exercise at an intensity of 40% to <60

TABLE 11.2 — CVD risk factors

MAJOR CARDIOVASCULAR DISEASE RISK FACTORS
1. **Age:** Men ≥ 45 yr, Women ≥ 55 yr.
2. **Diagnosed hypertension** (systolic ≥ 140 or diastolic ≥ 90 mmHg measured on two separate occasions) or on medications for treatment of hypertension.
3. **Dyslipidemia:** Serum LDL-Cholesterol ≥ 130 mg · dl^{-1} (≥3.37 mmol· L^{-1}), *or* HDL-Cholesterol < 40 mg · dl^{-1} (< 1.04 mmol· L^{-1}), *or* if no lipoproteins available use serum cholesterol ≥200 mg · dl^{-1} (≥5.18 mmol· L^{-1}), *or* on lipid lowering medications.
4. **Cigarette smoking.**
5. **Sedentary lifestyle.**
6. **Obesity:** BMI ≥ 30 kg · m^2 or waist girth >102 cm (40 inches) men, >88 cm (35 inches) women.
7. **Diabetes mellitus or prediabetes:** abnormal plasma glucose or glucose tolerance.
8. **Family history** of heart disease, heart attack, sudden death, coronary artery surgery, or other atherosclerotic disease prior to age 55 yr in father or other 1st degree male relative, or before age 65 yr in mother or other 1st degree female relative.
9. **High serum HDL-C Negative Risk Factor** if ≥60 mg · dl^{-1} (≥1.55 mmol· L^{-1})
SYMPTOMS SUGGESTIVE OF CARDIOVASCULAR OR METABOLIC DISEASE
1. Pain or discomfort in the chest or surrounding areas, including neck, arms, jaw, that appears to be related to cardiac ischemia.
2. Unusual shortness of breath with or without mild exertion.
3. Dizziness or syncope either during rest or exercise.
4. Difficulty breathing in supine position or sudden onset of labored breathing during sleep.
5. Ankle edema, especially bilateral at night.
6. Rapid or forceful heartbeats (palpitations) or unusually rapid, sustained heartbeat (tachycardia).
7. Intermittent claudication; pain in muscle, especially the legs, during exercise due to limited blood flow caused by diseased arteries.
8. Heart murmur may indicate serious heart valve problems or heart diseases which affect the flow of blood.
9. Feeling unusual tiredness, weakness, or fatigue, or easily short of breath when performing normal physical activities.

Major Cardiovascular Disease Risk Factors and Symptoms of Cardiopulmonary and Metabolic Disease Modified from Thompson, Gordon, and Pescatello, 2010.

percent of VO_2 reserve (VO_2R), and vigorous exercise as exercise at an intensity ≥60 percent of VO_2R. VO_2R is calculated as $[(VO_{2max}-VO_{2rest}) \times \%$ intensity desired$] + VO_{2rest}$ (Thompson, Gordon, and Pescatello, 2010). The accuracy of risk assessment and of the exercise prescription itself is much enhanced if physiologic and ECG data are available from a recent functional (or diagnostic) symptom-limited maximal graded exercise test (SLM-GXT) (procedures detailed in Laboratory 9 titled "Symptom-Limited Maximal Graded Exercise Test for Clinical and Sports Medicine Applications"). Thus, a SLM-GXT provides very valuable information important for designing the optimal exercise prescription, whether

TABLE 11.3 — Require—not require medical screening exam

Risk intensity	Low risk	Moderate Risk	High Risk
Moderate	no	no	yes
Vigorous	no	yes	yes

Recommendations for either **requiring** (=yes) or **not requiring** (=no) completion of a medical examination and diagnostic exercise test prior to participating in exercise based on risk categories for atherosclerotic cardiovascular disease Modified from Thompson, Gordon, and Pescatello, 2010.

or not it is required as a screening examination to ensure subject safety. As an alternative, some form of submaximal exercise testing can be employed to obtain data to improve the precision of the exercise prescription (see Laboratory 8 titled "Submaximal Exercise Testing for Estimating Aerobic Capacity").

Although it is common to emphasize cardiovascular training in prescribing exercise for health and fitness, a well-rounded exercise prescription should also include recommendations to improve muscle strength and flexibility. This is especially important with aging, since muscle mass is lost as part of the aging process. Exercise prescription for improving muscle strength and endurance will be detailed in Laboratory 12, "Exercise Prescription for Musculoskeletal Fitness and Health." In brief, it is generally recommended that at least one set of 8 to 12 repetitions of resistance training exercises that condition the major muscle groups of the body be performed at moderate intensity 2 days per week or more. Stretching exercises to improve or maintain optimal flexibility in a joint or series of joints should also be performed at least 3 times per week. The stretching exercises should consist a slow dynamic movement to the point of mild muscle or joint discomfort (stretched feeling), followed by a sustained, static stretch for 10 to 30 seconds; each exercise should be repeated 3 to 5 times. Often, the flexibility exercises are included in the warm-up/cool-down phases of the exercise session (Pollock et al., 2000; Thompson, Gordon, and Pescatello, 2010).

In addition to the assessment of health, risk status, and physical fitness, it is important to consider the factors described below in prescribing exercise. The prospect of success, as measured by the individual showing improvement in physical fitness and by adopting exercise as a life-long habit, is greatly improved when these factors are adequately addressed in the exercise prescription process.

1. Assess the individual's needs, interests, and objectives, and use this information to set realistic short-term and long-term goals.
2. Provide adequate education to the individual regarding such things as:
   a. Proper exercise clothing, footwear, and equipment.
   b. Principles of the exercise prescription process, including mode, frequency, intensity, duration, and progression.
   c. Physiologic changes that should be expected under normal circumstances during a single bout of exercise and in response to long-term exercise training.
   d. The methods of monitoring and recording physiologic responses to exercise, such as blood pressure and heart rate, and recognizing abnormal responses.
   e. Caution when exercising under adverse environmental conditions, such as heat and cold.
3. There is absolutely **NO** substitute for providing strong leadership and motivation to the individual, especially during the initial days and weeks of exercise training.

4. It is important for the person to understand that when exercising for health, the adage, "no pain, no gain" is **NOT** true. While well-trained athletes may need to push themselves through a pain threshold to realize maximum performance, severe pain associated with exercise training for the average fitness enthusiast is a telltale sign of overexertion. Continued exercise at this level of intensity may even be detrimental to long-term training. It is best to start slow and progress cautiously. Training sessions that are mildly fatiguing, but enjoyable, are best for promoting long-term adherence to a training program.
5. Fitness evaluations performed on a regular basis can serve an important role in motivation, and can be a source of positive reinforcement. However, if not understood properly in light of reasonable expectations for improvement, this same information could prove discouraging. It is important, therefore, that the individual be educated about the magnitude of fitness changes that are likely to occur in response to training. This, again, points to the importance of initially setting reasonable and achievable goals and objectives for the training program. In addition to the motivational role, these evaluations can also provide valuable information useful for making decisions about the progression rate of the exercise prescription.

Each exercise session should be carefully organized for efficiency. As a general rule, a single session should not last more than one hour. The organization for an ideal exercise session is shown in Table 11.4. Naturally, the session length and the length of time spent in each activity will need to be tailored to the individual's fitness goals and total time available for the exercise session. For example, some individuals may choose to place more emphasis on muscular training; consequently, more time will be spent on activities to promote muscle fitness and less on aerobic training. (See Laboratory 12, "Exercise Prescription for Musculoskeletal Fitness and Health"). Also, it is important to remember in designing exercise programs that, from a health perspective, it is much more critical that sedentary individuals be encouraged to "just do something physical" to increase their physical activity, even if they cannot commit a large block of time to exercise. Blair and associates (1992) have concluded that the greatest exercise benefit from the standpoint of disease prevention is found in previously sedentary individuals who become moderately active. This is the challenge to the exercise professional, to develop an exercise prescription that meets the goals and objectives of the individual within the constraints of their current health and fitness status and their limitations, including the limitation of time available to exercise.

This laboratory is designed to lead the student through the general exercise prescription process for aerobic fitness, providing the guiding principles in a reasonable model to follow in designing exercise prescriptions for the apparently healthy. It is important to note that the specific recommendations given are not intended to be applied in all circumstances. The prescription process is, by definition "individualized," and, therefore, variations in the specific application of the model components are common. Through years of experience, the exercise professional will develop the "art" of providing safe and effective exercise prescriptions for individuals of varying health and fitness status.

Equipment Needed

Stopwatch

Treadmill

TECHNIQUES AND PROCEDURES

The underlying assumptions for this laboratory exercise are that: (1) the procedures will be applied in an exercise program where the majority of prospective participants are apparently healthy adults; (2) the

TABLE 11.4 — Organization phases of workout

Session Phase	Time (min)	Purpose and Examples of Exercises
Warm-up	5–10	1. Gradually increase metabolic rate from resting to that required by conditioning phase of exercise session. 2. Static and dynamic stretching, light to moderate calisthenics and muscular endurance exercises. 3. Light (<40% VO_2R) to moderate (between 40% and 60% VO_2R) cardiovascular endurance activities.
Conditioning	20–60	1. Mix of aerobic and resistance exercise dependent on objectives of physical training, day of week, period of training. May be sport activities. 2. Promote increase in aerobic working capacity (VO_{2max}). Continuous or discontinuous (10 min minimal per session to accumulate 20 to 60 min $\cdot$ d^{-1}) exercise involving large muscle groups producing heart rates of the prescribed intensity such as; fast walking, jogging, bicycling, swimming, skating, dancing, and vigorous games. 3. Increase muscular strength, endurance, and tonus of all major muscle groups. Heavy calisthenics, weight training, circuit weights.
Cool-down	5–15	1. Gradually decrease metabolic rate for recovery. 2. Light (<40% VO_2R) to moderate (between 40% and 60% VO_2R) cardiovascular and muscular endurance exercise. 3. Stretching and relaxation exercises.

Three Phases of an Ideal Exercise Session Adapted from Thompson, Gordon, and Pescatello, 2010.

primary goal of the subject will be to improve health-related fitness, specifically aerobic fitness; (3) the student has completed Laboratory 9 titled "Symptom-Limited Maximal Graded Exercise Test for Clinical and Sports Medicine Applications," and the functional data are available from which to develop the exercise prescription (if these data are not available, Table 11.5 contains sample SLM-GXT data which can be used for this laboratory experience); and (4) the subject is currently sedentary and desires to begin a moderate exercise program. Although in large group settings it is not always possible to apply all aspects of individualized exercise prescription, it is essential that all participants be screened for risk factors, disease, and exercise contraindications, and that they be given duration and heart-rate intensity guidelines that take into account age, gender, fitness level, and personal goals.

PREPARTICIPATION SCREENING

1. As with any exercise test or exercise prescription, all subjects should be adequately informed about the attendant risks and sign an informed consent prior to participating in an exercise program. An informed consent document easily adaptable for the exercise prescription process can be found in the Laboratory 8 entitled "Submaximal Exercise Testing for Estimating Aerobic Capacity." Once the informed consent has been signed and witnessed, proceed with the pre-participation screening.

TABLE 11.5 — Results of SLM-GXT for exercise prescription

PRE-EXERCISE DATA				
Name: Jane Doe	Age: 24	Sex: Female	Height: 163 cm	Weight: 50.9 kg
Medications: None	Resting Heart Rate: 61 bpm		Resting Blood Pressure: 110/60	
SLM-GXT ACHIEVED VALUES				
Protocol: Bruce		Max Heart Rate: 186		Max Blood Pressure: 178/48
Max Rate Pressure Product: 331		Max Treadmill Time: 13:34		RPE: 18
VO_{2max}: 49.1 ml $\cdot$ kg$^{-1}\cdot$ min^{-1}		Max METS: 14.0		Functional Aerobic Impairment: −45
Limiting Symptoms: General Fatigue				
GENERAL INTERPRETATION				
ECG and SLM-GXT normal				

Result from a SLM-GXT as an example of the data which can be used in developing the exercise prescription.

2. Ask the subject to complete the Personal Information, Health Information, and Goals and Exercise Preference sections of the Exercise Prescription Subject Questionnaire.
3. Review the Health Information section of the Exercise Prescription Subject Questionnaire (refer to Table 11.2 in evaluating this information), then complete the Health Information and Classification section of the Exercise Prescription Worksheet.
4. **Decision Point 1.** Review the description of the three health categories described previously to assign the subject to the appropriate category: (1) Low risk, (2) Moderate Risk, or (3) High Risk.
5. **Decision Point 2.** Review the ACSM recommendations presented in Table 11.3. Indicate whether or not this subject should be required to complete a medical exam and diagnostic exercise test prior to participation in the exercise program. Also indicate if they are cleared for participation in a moderate or vigorous exercise program. (Assume that this subject wishes to begin a *moderate* exercise program.) If SLM-GXT data are not available, use the data in Table 11.5 to gain experience prescribing exercise.
    a. **YES**, medical clearance is recommended. An exercise prescription should not be given until the subject is properly evaluated by a physician and has completed a diagnostic exercise test.
    b. **NO**, medical clearance is not required. Proceed through the exercise prescription process.
6. Complete the Resting and Available SLM-GXT Data section of the Exercise Prescription Worksheet. Exercise test data in Table 11.5 may be used for this step if other SLM-GXT data are not available.
7. Using Table 11.6, determine the fitness category and record in the appropriate blank space on the worksheet. An alternate strategy to classify subjects into fitness categories to assist in determining the initial exercise prescription components has also been published by the American College of Sports Medicine (Thompson, Gordon, and Pescatello, 2010). The fitness category in Table 11.6 assumes VO_{2max} has

recently been measured from SLM-GXT or submaximal exercise testing. If VO_{2max} from exercise testing is not available, it can be reasonably estimated using the NASA/JSC physical activity scale (PA-R), age (yr), gender (F = female, M = male), and body mass index (BMI) as follows (Jackson et al., 1990):

$$VO_{2max} = 56.363 + 1.92(PA\text{-}R) - 0.381(Age) - 0.754(BMI) + 10.987(F=0, M=1)$$

The PA-R can be found at:

http://www.ohp.nasa.gov/disciplines/hpromo/hpwTeam/meetings/handouts/2006-03_NASA_PAScale.pdf

   a. Note that Table 11.6 also contains information related to intensity, duration, and frequency for the initial conditioning stage of the exercise prescription, which will be used in the subsequent section to develop the exercise prescription.

DEVELOPING THE EXERCISE PRESCRIPTION

1. Complete the Beginning Exercise Prescription section of the Exercise Prescription Worksheet. Begin with the selection of the recommended exercise **Mode**.

   a. The mode should be selected in accordance with the individual goals, likes and dislikes as detailed on the Exercise Prescription Subject Questionnaire. The mode for aerobic training may be any activity that is rhythmic, uses large muscle groups, and can be maintained for the prescribed duration. Examples of aerobic activities include walking, jogging, running, swimming, bicycling, rowing, cross-country skiing, skating, dancing, stair-climbing, and endurance game activities.

2. Determine the **Frequency** of each exercise session. This is generally given as the number of exercise sessions per week. However, when an individual cannot complete the duration requirements in a single session either due to physical or time limitations, shorter duration sessions, as short as 10 minutes each, may be completed throughout each day. For example, if the duration requirements were for 20 minutes per day, this could be divided into two 10-minute sessions. Suggested guidelines for the beginning exercise frequency are given in Table 11.6.

TABLE 11.6 — Fitness categories for the exercise prescription

Fitness Category	VO_{2max} (ml · kg^{-1} · min^{-1})	Intensity (% VO_2R)	Duration (Minutes)	Frequency (times·wk^{-1})
1	7–17	40–50	15	3
2	17.5–27.5	45–55	15	3
3	28–38	50–60	15–20	3
4	38.5–45	55–65	15–20	3–4
5	45.5–52	60–70	20–25	3–4
6	52.5–59	65–75	25–30	3–5
7	59.5–69.5	70–80	30–40	3–5
8	≥70	75–85	40–60	4–5

Fitness Categories Based on Achieved VO_{2max} and Recommended Beginning Intensity, Duration, and Frequency Ranges for the Initial Conditioning Stage of the Exercise Prescription.
VO_2R is calculated as [($VO_{2max} - VO_{2rest}$) × % intensity desired] + VO_{2rest}.
Note: Fitness categories modified from Pollock and Wilmore, 1990.

3. Calculate the exercise **Intensity** for the aerobic training portion of the exercise session.
   a. First determine the beginning intensity range for the initial conditioning stage of the exercise prescription from Table 11.6 and record this value in the %VO_2R column. As a reminder, VO_2R is calculated as $[(VO_{2max} - VO_{2rest}) \times \%$ intensity desired$] + VO_{2rest}$ (Thompson, Gordon, and Pescatello, 2010).
   b. Since the heart rate is a good indicator of the physiologic stress experience by the body during exercise, the most common method of prescribing and monitoring exercise intensity is to calculate a *target heart rate range.* This range corresponds to the desired intensity range for exercise training. To calculate the target heart rate range (THR), use the beginning intensity range from Table 11.6 and follow this sequence of steps.
      i. First, calculate the Heart Rate Reserve (HRR) (bpm).

      HRR (bpm) = HR_{max} (bpm) − HR_{rest} (bpm)
      Where:

      HR_{max} (bpm) = maximal heart rate achieved during a SLM-GXT.

      When not available, HR_{max} (bpm) can be estimated as:

      207 − (0.7× subject age) (Gellish et al., 2007)

      or 220 − subject age

      HR_{rest} (bpm) is the heart rate measured with the subject resting comfortably, preferably upon waking in the morning. In practice, the resting, supine heart rate from the graded exercise test is often used.

      ii. Next, calculate the **Lower Limit** and **Upper Limit** heart rate for the **Target Heart Rate Range**.

      Lower Limit (bpm) = [(Lowest Intensity % ÷ 100) × HRR] + HR_{rest} (bpm)
      Upper Limit (bpm) = [(Highest Intensity % ÷ 100) × HRR] + HR_{rest} (bpm)
      Where:

      Lowest Intensity % = the percent of maximal effort corresponding to the **lowest** exercise intensity desired for training.

      Highest Intensity % = the percent of maximal effort corresponding to the **highest** exercise intensity desired for training.

      HRR = the heart rate reserve calculated previously.

      HR_{rest} = resting heart rate as defined above.

   c. Change the values of the target heart rate range from beats per minute (bpm) to beats per 10 seconds by dividing the lower limit and upper limit heart rates by a factor of 6. It is common to express the exercise heart rate as beats per 10 seconds (or sometimes per 15 seconds) because it is much more efficient to count the pulse rate for 10 seconds than for a complete minute during exercise. Record the target heart rate range as a 10-second value in the THR column on the worksheet.
      i. If a heart rate watch monitor is available to be worn by the subject during exercise to monitor heart rate, this step will not be required, and the target heart rate range should be recorded in bpm values.

d. Calculate the exercise intensity MET range, and record this range in the column titled METS. The MET range for exercise prescription can be calculated in a manner similar to calculating the target heart rate range by using the following equations.

Lower MET Limit = Lowest Intensity % × Maximum METS

Upper MET Limit = Highest Intensity % × Maximum METS

Where:

> Lowest Intensity % is the percent of maximal effort corresponding to the **lowest** exercise intensity desired for training.
>
> Highest Intensity % is the percent of maximal effort corresponding to the **highest** exercise intensity desired for training.
>
> Maximum METS is the value corresponding to the maximal working capacity of the individual measured on an exercise test.

Once the MET intensity range is known for an individual, physical activities with energy expenditures in the prescription range can be recommended. For example, playing golf (4–7 MET activity) might be recommended for a person with a prescribed MET range of 6–7, but playing racquetball (8–12 MET activity) would not. The use of the MET range for exercise prescription is particularly useful when the subject desires to participate in game activities in which the use of heart rate to monitor exercise intensity may not be feasible. This method is also used in clinical settings with cardiac or pulmonary rehabilitation patients. In this case, physical activities can be prescribed which do not exceed the safe exertional limits of the patient. The MET and energy expenditure values of common leisure activities are published in several sources (Heyward, 1998; Thompson, Gordon, and Pescatello, 2010). When prescribing intensity using METS or kilocalories, it is important to recognize that environmental factors can dramatically impact the energy cost of the selected activities. Changes in wind velocity, heat, humidity, or altitude can alter the workload imposed on the body, and thereby affect the relative intensity of the exercise activity. To ensure that a safe intensity is maintained irrespective of changing environmental conditions, heart rate recommendations should always accompany MET prescriptions.

e. Record the recommended RPE range in the intensity section of the Exercise Prescription Worksheet. Along with a target heart rate and MET prescription, it is common to provide a Rating of Perceived Exertion (RPE) (Borg, 1982) corresponding to the prescribed exercise intensity as part of the exercise prescription. RPE scales are included in the Laboratory 9 entitled "Symptom-Limited Maximal Graded Exercise Test for Clinical and Sport Medicine Applications." Using the 15-point scale, an RPE of 12 to 13 generally corresponds to a target heart rate of about 60%, and 16 corresponds to 85%.

   i. Whenever possible the RPE and heart rate data from an actual SLM-GXT should be used in prescribing the RPE intensity. Then the individual's true RPE and heart rate relationship will be known, and the RPE corresponding to the prescribed heart rate can be determined with a high degree of precision.

   ii. In general, during the initial phase of training, an RPE of 11 to 13 is appropriate. The prescribed RPE should follow a natural progression corresponding to the increase in the target heart rate as the training intensity is progressively increased over time.

   iii. Along with the RPE, the "talk test." can be used as an informal assessment of intensity during exercise. In this case, when an individual is too winded to maintain a normal conversation with a partner while exercising, they are most likely working at an intensity that is too high.

4. Determine the **Duration** for conditioning phase of each exercise session exclusive of the warm-up and cool-down, and record this value in the appropriate column on the worksheet. Duration should generally range from 20 to 60 minutes depending upon the subject's current fitness level. Duration guidelines for beginning exercise, based on the subject's fitness category, are given in Table 11.6.

   a. It is common during the initial and improvement conditioning stages of the exercise program to intersperse a rest/recovery interval between periods of exercise at the prescribed intensity; that is, aerobic conditioning may be discontinuous, and the exercise prescription resembles that of interval training. For example, if the total duration is prescribed to be 20 minutes, this total time may be completed in 4 bouts of 5 minutes of exercise at the prescribed intensity, interspersed with 2 minutes of recovery at a lower intensity. Thus, an individual who would not be able to complete 20 minutes of continuous exercise can complete the prescribed duration when broken up into smaller intervals. This practice can also help prevent injuries and reduce the muscle soreness that often accompanies the early stages of exercise conditioning.

5. When possible, estimate the beginning workload or pace required to produce the desired intensity. This can be accomplished for a number of aerobic activities by applying the metabolic equations shown in Table 11.7.

   a. For example, suppose a subject weighing 70 kg chooses a stationary cycle ergometer as the mode of exercise, and their exercise prescription calls for a 6–8 MET range. The work rate needed to produce this intensity can be estimated by applying the following sequence of equations.

      i. First, convert the prescription MET range to VO_2 (ml · kg^{-1} · min^{-1}).
         1. Lowest VO_2 (ml · kg^{-1} · min^{-1}) = 6 METS × 3.5 ml · kg^{-1} · min^{-1} = 21 ml · kg^{-1} · min^{-1}
         2. Highest VO_2 (ml · kg^{-1} · min^{-1}) = 8 METS × 3.5 ml · kg^{-1} · min^{-1} = 28 ml · kg^{-1} · min^{-1}

      ii. Next, chose the proper equation for the selected mode from Table 11.7, and rearrange the equation to solve for work rate or pace. In this example, the equation for leg ergometer work would be selected, and the equation rearranged to solve for work rate in kpm · min^{-1} is as follows.
         1. kpm · min^{-1} = (VO_2 − 7 ml · kg^{-1} · min^{-1}) × kg body mass ÷ 1.8 ml O_2 · kpm^{-1}

      iii. Substitute the lower and upper VO_2 limits calculated above in step i above and the individual's body weight in kg into the equation derived in step ii, then solve for the lower and upper limit work rate. For this example for a 70 kg subject, the cycle ergometer workloads that would result in the prescribed exercise VO_2 can be calculated as follows:
         1. **Lower Limit Workload** (kpm · min^{-1}) = (21.0 − 7.0) × 70 ÷ 1.8 = 544.4 ≈ 550
         2. **Upper Limit Workload** (kpm · min^{-1}) = (28.0 − 7.0) × 70 ÷ 1.8 = 816.7 ≈ 820

   b. Thus, in this example, the beginning workload would be set between 550 and 820 kpm· min^{-1}. Since the conversion factor for watts is 1 W = 6.1 kpm· min^{-1} the work rate range in watts would be about 90 and 135 W. Similar calculations can be made to estimate the beginning pace for walking, jogging, or stepping by applying the respective equations shown in Table 11.7. Record the beginning pace or workload in the appropriate column on the Exercise Prescription Worksheet.

6. To test the accuracy of the beginning exercise prescription, ask the subject to exercise for at least 3 minutes after a warm-up at the specified workload or pace. Check the heart rate by palpating the pulse after 3 minutes of exercise or by reading the heart rate displayed by the heart rate watch monitor, and compare to the target heart rate range. If the heart rate is too low, increase the workload or pace; decrease the workload or pace if the heart rate is too high. Check the heart rate again after another 3 minutes if adjustments were made. Continue this process until a work rate or pace is found

TABLE 11.7 — Metabolic formulas for VO_2

Activity	Calculation of Oxygen Uptake (VO_2) in $ml \cdot kg^{-1} \cdot min^{-1}$
Treadmill Walking*	= (Speed * 0.1) + (1.8 *Speed* fractional grade) + 3.5
Treadmill Running*	= (Speed * 0.2) + (0.9*Speed *fractional grade) + 3.5
Leg Ergometer**	= (Work Rate * 1.8 ÷ Body Mass) + 7
Arm Ergometer**	= (Work Rate * 3 ÷ Body Mass) + 3.5
Stepping§	= (Step Rate *0.2)+(Step Height * Step Rate*1.8*1.33) + 3.5
Common Conversion Factors for Exercise Prescription	
1 MET = $VO_{2(ml/min)}$ ÷ body $wt_{(kg)}$ ÷ 3.5($mlO_2 \cdot kg^{-1} \cdot min^{-1} \cdot MET^{-1}$) OR = VO_2 ($ml \cdot kg^{-1} \cdot min^{-1}$) ÷ 3.5($mlO_2 \cdot kg^{-1} \cdot min^{-1} \cdot MET^{-1}$) Estimated Energy Expenditure ($kcal \cdot min^{-1}$) = METs * Body $wt_{(kg)}$ * .0175($kcal \cdot kg^{-1} \cdot min^{-1} \cdot MET^{-1}$) Estimated Energy Expenditure ($kcal \cdot min^{-1}$) = $VO_2(L \cdot min^{-1})$ * 5($kcal \cdot LO_2^{-1}$) 1 $mi \cdot h^{-1}$ = 26.8 $m \cdot min^{-1}$ 1 kg = 2.2 lbs 1 watt = 6.1 $kpm \cdot min^{-1}$ = 0.01433 $kcal \cdot min^{-1}$ = 1 joule $\cdot sec^{-1}$ 1 kpm = 9.807 joules = 0.0023 kcal 1 Joule (J) = 1 Newton meter (Nm)	

* Treadmill walking speeds 50 to 100 m/min (1.9 to 3.7 mph). Running speeds >134 m/min (>5 mph) and slower speeds of 80–134 m/min (3–5 mph) if truly running. Unit of measure for speed is $m \cdot min^{-1}$, fractional grade is decimal equivalent of percent grade.
** Work rate for leg and arm ergometer is $kpm \cdot min^{-1}$, body mass is kg.
§ Unit for step height is meters (m), step rate is $steps \cdot min^{-1}$.
Common Conversion Factors and Metabolic Formulas for Determining Oxygen Uptake (VO_2 $ml \cdot kg^{-1} \cdot min^{-1}$) from Various Aerobic Activities Modified from Thompson, Gordon, and Pescatello, 2010.

that produces a heart rate in the prescribed range. Also, verify the accuracy of the RPE prescription during this time, and adjust accordingly to correspond to the prescribed heart rate. Once the proper work rate or pace is found, this becomes the recommendation for the first few exercise sessions of the initial conditioning stage.

7. Provide general recommendations to the subject for the organization of a typical exercise session by completing the section titled Phase Organization of Typical Exercise Session on the Exercise Prescription Worksheet.

8. Make recommendations for the **Progression** of exercise by completing the sections of the Exercise Prescription Worksheet titled Initial Conditioning Stage, Improvement Conditioning Stage, and Maintenance Conditioning Stage. It is important to recognize that no recommendations can be dogmatically applied, since progression will be individualized, and depends upon the subject's initial fitness category, general health, and exercise goals. Furthermore, the initial recommendations often need to be altered after beginning a training program for a variety of reasons; for example, to account for unforeseen changes in the health status, training rate, or injury status of the subject. An example of an exercise prescription is given in Table 11.8. General considerations for each conditioning stage are as follows.

    a. Initial Conditioning Stage
        i. Low to moderate intensity and duration should be used. One should consider beginning with a discontinuous approach to the exercise prescription especially with those who are very unfit, to minimize delayed onset muscle soreness.

ii. Be sure to include adequate stretching and light calisthenics during each exercise session, which will also help to limit soreness and help prevent injury.
iii. Encourage the individual to engage in low impact activities.
iv. This stage should normally last 2 to 6 weeks depending on the individual's initial fitness level and training response. Those who are initially unfit will naturally spend more time in this stage than those who are very fit.
v. Prescribe small increments in the intensity/duration progression.

b. Improvement Conditioning Stage
  i. The duration and intensity should be increased regularly every one to three weeks. As a general rule, both of these factors should not be increased simultaneously. Move from discontinuous to continuous exercise.
  ii. The rate of progression can be more rapid than during the initial conditioning stage. The rate will be dictated by individual training response.

TABLE 11.8 — Example Ex Rx progression

Stage of Ex Rx	Week	Frequency (d·wk^{-1})	Intensity (% VO$_2$R)	Total Duration (min)	Interval (min)	Recovery (min)	Repeats
Initial Conditioning Stage	1	3	50–60	16	2	1	8
	2	3	50–60	18	3	1	6
	3	3	55–65	18	3	1	6
	4	3	55–65	20	4	1	5
	5	3	60–70	20	5	1	4
	6	3	65–75	20	5	1	4
Improvement Conditioning Stage	7–8	3	65–75	20	10	3	2
	9–10	3	65–75	24	12	3	2
	11–12	3	65–75	28	14	2	2
	12–13	3	65–75	30	15	1	2
	14–15	3	65–75	30	continuous	-	-
	16–17	3–4	70–80	30	continuous	-	-
	18–19	3–4	75–85	30	continuous	-	-
	20–21	3–4	75–85	32	continuous	-	-
	22–23	3–5	75–85	34	continuous	-	-
	24–25	3–5	75–85	36	continuous	-	-
	26–27	3–5	75–85	40	continuous	-	-
Maintenance Conditioning Stage	27+	3–5	75–85	40+	continuous		

Example of progression through the stages of exercise prescription for a relatively unfit person.

iii. Age must be considered in determining the rate of progression. Estimates are that it takes about 40% longer to adapt to training for each decade in life after age 30.
iv. This stage normally lasts 4 to 5 months.

c. Maintenance Conditioning Stage
   i. This stage commences when the individual's fitness goal has been reached. The objective in this stage is to maintain the fitness level attained through training.
   ii. Continue the workout schedule, which will maintain conditioning level.

QUESTIONS AND ACTIVITIES

1. Assume that the individual for whom you provided an exercise prescription in this laboratory will be traveling from a cool, dry climate to a hot, humid climate. Discuss any changes you would recommend in their exercise prescription, and provide rationale for your recommendations.
2. Why might heart rate be a more useful measure of the exercise intensity for a jogger than a mile pace?
3. How would you alter your exercise prescription for a subject who was obese?
4. Discuss ways in which you would account for individual differences in fitness in prescribing exercise to individuals in a dance aerobics class.
5. Compare and contrast the general exercise prescription you would give to a healthy 37-year-old woman with a VO_{2max} of 56 ml · kg^{-1} · min^{-1} with another woman of the same age having a VO_{2max} of 28. Assume both women completed an uneventful SLM-GXT, and both are free of any known diseases.

QUESTIONS AND ACTIVITIES FOR GOING FURTHER

1. Assume you are prescribing aerobic exercise to a 55-year-old women with hypertension treated by daily doses of propranolol. What data would you want before prescribing exercise for this individual and how might the ingestion of this medication affect the exercise prescription you would give them?
2. Assume your newest client is a 35-year-old male subject who weighs 100 kg, is 178 cm tall, and expresses a primary goal of exercise training to lose 20 pounds of body weight. Data from the SLM-GXT shows he is healthy and free of contraindications for exercise, has a normal ECG response to exercise, and achieved a $HR_{max} = 188$, $VO_{2max} = 35$ ml · kg^{-1} · min^{-1}, and normal systolic and diastolic blood pressure. His current diet is calorically neutral, that is, energy intake balances his daily energy expenditure, and he vows not to change his diet during exercise training. Write an exercise prescription for this subject showing the caloric expenditure during each training session, and the projected weight loss each week until he achieves his weight loss goal. Consider that he will complete a second SLM-GXT after the eighth week of the exercise program, and that his VO_{2max} will have improved 15%.
3. Assume you are in the eighth week of training a 44-year-old man. His initial VO_{2max} placed him in fitness category 4 when he began his exercise program, and he has responded well to training. He lives and has been training at 300 feet above sea level, and will be traveling for 3 weeks to an elevation of 9,000 to 11,000 feet. What exercise prescription would you give him while he is traveling at altitude? Give physiologic reasons for your new exercise prescription.

REFERENCES

Blair, S. N., H. W. Kohl, N. F. Gordon, and R. S. Paffenbarger. "How Much Physical-Activity Is Good for Health." *Annu. Rev. Public Health* 13 (1992): 99–126.

Borg, GAV. "Psychophysical Bases of Perceived Exertion." *Med. Sci. Sports Exerc.* 14.5 (1982): 377–81.

Centers for Disease Control and Prevention. "Trends in Leisure-Time Physical Inactivity by Age, Sex, and Race/Ethnicity, United States, 1994–2004." *MMWR Morb Mortal Wkly Rep* 54.39 (2005): 991–94.

Crouse, S. F., B. C. O'Brien, P. W. Grandjean, R. C. Lowe, J. J. Rohack, and J. S. Green. "Effects of Training and a Single Session of Exercise on Lipids and Apolipoproteins in Hypercholesterolemic Men." *J. Appl. Physiol.* 83.6 (1997a): 2019–28.

Crouse, S. F., B. C. O'Brien, P. W. Grandjean, R. C. Lowe, J. J. Rohack, J. S. Green, and H. Tolson. "Training Intensity, Blood Lipids, and Apolipoproteins in Men with High Cholesterol." *J. Appl. Physiol.* 82.1 (1997b): 270–77.

Debusk, R. F., U. Stenestrand, M. Sheehan, and W. L. Haskell. "Training Effects of Long Versus Short Bouts of Exercise in Healthy-Subjects." *Am. J. Cardiol.* 65.15 (1990): 1010–13.

Fletcher, G. F., S. N. Blair, J. Blumenthal, C. Caspersen, B. Chaitman, S. Epstein, H. Falls, E. S. S. Froelicher, V. F. Froelicher, and I. L. Pina. "Statement on Exercise: Benefits and Recommendations for Physical-Activity Programs for all Americans—A Statement for Health-Professionals by the Committee on Exercise and Cardiac Rehabilitation of the Council on Clinical Cardiology, American Heart Association." *Circulation* 86.1 (1992): 340–44.

Gellish, R. L., B. R. Goslin, R. E. Olson, A. McDonald, G. d. Russi, and V. K. Moudgil. "Longitudinal Modeling of the Relationship Between Age and Maximal Heart Rate." *Med. Sci. Sports Exerc.* 39.5 (2007): 822–29.

Hagberg, J. M., J. J. Park, and M. D. Brown. "The Role of Exercise Training in the Treatment of Hypertension: An Update." *Sports Med.* 30.3 (2000): 193–206.

Haskell, W. L., I. Lee, R. R. Pate, K. E. Powell, S. N. Blair, B. A. Franklin, C. A. Macera, G. W. Heath, P. D. Thompson, and A. Bauman. "Physical Activity and Public Health: Updated Recommendation for Adults from the American College of Sports Medicine and the American Heart Association." *Circulation* 116.9 (2007): 1081–93.

Heyward, V. H. *Advanced Fitness Assessment & Exercise Prescription.* Champaign, IL: Human Kinetics, 1998.

Jackson, A. S., S. N. Blair, M. T. Mahar, L. T. Wier, R. M. Ross, and J. E. Stuteville. "Prediction of Functional Aerobic Capacity without Exercise Testing." *Med. Sci. Sports Exerc.* 22.6 (1990): 863–70.

Lee, I. "Physical Activity and Cardiac Protection." Current Sports Medicine Reports 9.4 (2010): 214–19.

Pate, R. R., M. Pratt, S. N. Blair, W. L. Haskell, C. A. Macera, C. Bouchard, D. Buchner, W. Ettinger, G. W. Heath, A. C. King, A. Kriska, A. S. Leon, B. H. Marcus, J. Morris, R. S. Paffenbarger, K. Patrick, M. L. Pollock, J. M. Rippe, J. Sallis, and J. H. Wilmore. "Physical-Activity and Public Health: A Recommendation from the Centers for Disease Control and Prevention and the American College of Sports Medicine." 273.5 (1995): 402–07.

Pollock, M. L., and J. H. Wilmore. *Exercise in Health and Disease: Evaluation and Prescription for Prevention and Rehabilitation.* Philadelphia: W.B. Saunders, 1990.

Pollock, M. L., B. A. Franklin, G. J. Balady, B. L. Chaitman, J. L. Fleg, B. Fletcher, M. Limacher, I. L. Pina, R. A. Stein, M. Williams, and T. Bazzarre. "Resistance Exercise in Individuals with and without Cardiovascular Disease: Benefits, Rationale, Safety, and Prescription—An Advisory from the Committee on Exercise, Rehabilitation, and Prevention, Council on Clinical Cardiology, American Heart Association." *Circulation* 101.7 (2000): 828–33.

Rogers, C.J., L. H. Colbert, J. W. Greiner, S. N. Perkins, and S. D. Hursting. "Physical Activity and Cancer Prevention: Pathways and Targets for Intervention." Sports Med. 38.4 (2008): 271–96.

Sanz, C., J. Gautier, and H. Hanaire. "Physical Exercise for the Prevention and Treatment of Type 2 diabetes." *Diabetes Metab.* 36.5 (2010): 346–51.

Thompson, P.D., S. F. Crouse, B. Goodpaster, D. Kelley, N. Moyna, and L. Pescatello. "The Acute Versus the Chronic Response to Exercise." *Med. Sci. Sports Exerc.* 33.6 (2001): Suppl S438–45; discussion S452–3.

Thompson, P. D., B. A. Franklin, G. J. Balady, S. N. Blair, D. Corrado, M. Estes, J. E. Fulton, N. F. Gordon, W. L. Haskell, M. S. Link, B. J. Maron, M. A. Mittleman, A. Pelliccia, N. K. Wenger, S. N. Willich, and F. Costa. "Exercise and Acute Cardiovascular Events: Placing the Risks into Perspective." *Med. Sci. Sports Exerc.* 39.5 (2007): 886–97.

Thompson, W. R., N. F. Gordon, and L. S. Pescatello, eds. *ACSM's Guidelines for Exercise Testing and Prescription,* Philadelphia: Lippincott Williams & Wilkins, 2010.

Weinheimer, E. M., L. P. Sands, and W. W. Campbell. "A Systematic Review of the Separate and Combined Effects of Energy Restriction and Exercise on Fat-Free Mass in Middle-Aged and Older Adults: Implications for Sarcopenic Obesity." *Nutr. Rev.* 68.7 (2010): 375–88.

Wenger, H. A., and G. J. Bell. "The Interactions of Intensity, Frequency and Duration of Exercise Training in Altering Cardiorespiratory Fitness." *Sports Med.* 3.5 (1986): 346–56.

EXERCISE PRESCRIPTION SUBJECT QUESTIONNAIRE

(Modified from Thompson, Gordon, and Pescatello, *ACSM's Guidelines for Exercise Testing and Prescription,* Philadelphia: Lippincott Williams & Wilkins, 2010.)

Personal Information

Name _____ Date ___/___/___ Phone _____

Address _____

Sex (M or F) _____ Height _____ in Weight _____ lb Age _____ yr

Health Information Please check ONLY those that apply to you.

___ 1. My doctor has told me that I have heart problems, and I should only do the physical activities recommended by my doctor.

___ 2. I often feel pains in my chest, or have abnormal heart beats when I work or exercise.

___ 3. In the past month I have had chest pains when I was not doing any physical work or exercise.

___ 4. I have times when I feel faint, dizzy, or even passed out, or I have lost my balance when I became dizzy.

___ 5. My doctor has told me that I have high blood pressure or high cholesterol.

___ 6. I have bone or joint problems (e.g., back, knee, or ankle) that keep me from exercising or might be made worse if I exercise.

___ 7. I do not regularly do physical activity hard enough to make me sweat.

___ 8. I regularly take medications prescribed by my doctor.

    a. Give purpose: _____

    b. Name/Dose: _____

___ 9. I smoke or use tobacco products regularly.

___ 10. Women: I am now or might be pregnant.

___ 11. There are other reasons why I think I should not exercise or be careful exercising.

    a. List reasons: _____

NOTE: If any items are checked, the exercise prescription should be postponed until further review by a qualified professional. Medical clearance may be necessary before the individual begins to exercise or takes part in an exercise test.

Personal Exercise Goals and Preferences

1. In this section, please explain your personal goals for your exercise program? What are the main things you hope to gain by participating in regular physical activities? These may be things like: improve my overall health, increase my endurance, lose weight, meet new people, gain muscle strength, strengthen my bones, train for a specific sport or activity, etc.

   _____
   _____
   _____
   _____
   _____
   _____

2. What are the types of exercises and physical activities you would most like to do? List as many as you wish. These may be things like lift weights, walk, bicycle, jog, a sport, etc.

   _____
   _____
   _____
   _____
   _____
   _____

EXERCISE PRESCRIPTION WORKSHEET

Pre-Prescription Information

Name _____ Age _____ Sex _____ Date ___/___/___

Height _____ in   Weight _____ lb

Health Information and Classification	Resting and Available SLM-GXT Data
CVD Risk Factors _____	Resting Heart Rate _____
Symptoms _____	Resting Blood Pressure _____/_____
Activity Habits _____	Max Heart Rate _____
Conditions of Concern _____	Max Blood Pressure _____/_____
Medications _____	Max Rate Pressure Product _____
Risk Classification: _____ Low	Max Treadmill Time ____ min ____ sec
_____ Moderate	Max RPE _____ VO_{2max} _____
_____ High	Max METS _____ FAI _____
Refer for Medical Clearance? ___ Yes	Limiting symptoms _____
___ No	GXT Interpretation _____
Cleared for: ___ Moderate Exercise	Fitness Category _____
___ Vigorous Exercise	

Beginning Exercise Prescription

Mode	Frequency (d · wk^{-1})	Intensity Range				Duration (min)	Pace or Load
		%VO$_2$R	THR	METS	RPE		

Phase Organization of Typical Exercise Session		
Session Phase	Activity Time (min)	Mode or Activities Types
Warm-up		
Conditioning		
Cool-down		

EXERCISE PRESCRIPTION WORKSHEET

Initial Conditioning Stage								
Week	Frequency (d × wk-1)	Intensity			Total Duration (min)	Exercise or Interval (min)	Recovery (min)	Repeats
		%VO2R	THR	METS				
Improvement Conditioning Stage								
Maintenance Conditioning Stage								

CHAPTER 12
EXERCISE PRESCRIPTION FOR MUSCULOSKELETAL FITNESS AND HEALTH

OUTCOME OBJECTIVES

Following successful completion of this exercise, the student should be able to:

1. Evaluate needs, current health and fitness status, and goals for setting up an exercise prescription for musculoskeletal health for apparently healthy individuals.
2. Design an individualized program for musculoskeletal fitness.
3. Make recommendations for the design of a resistance fitness program as regards warm-up, exercise, and cool-down.

INTRODUCTION

An increased emphasis has been placed on musculoskeletal health and fitness in clinical settings within the past 10 to 15 years, and the American College of Sports Medicine (ACSM) strongly endorses the inclusion of resistance training, sometimes called weight training or strength training, in the exercise prescription process for adults (Ratamess et al., 2009). Furthermore, resistance exercise is considered an important component of training patients with known cardiovascular disease (Pollock et al., 2000). While cardiovascular complications are still the major cause of death among adult populations, aging is also associated with a decreased ability to perform activities of daily living (ADL) resulting in a lower quality of life. Many of these activities require at least a moderate amount of musculoskeletal fitness. Additionally, osteoporosis affects a great many postmenopausal women (age 50–55+ years) as well as elderly men (75–80+ years). Improving musculoskeletal fitness both helps the elderly population in their ability to maintain ADL and to delay the onset and severity of osteoporosis (ACSM, 2004). It has also been shown to aid in managing obesity, Type II diabetes, and osteoarthritis (Nelson et al., 2007). Table 12.1 shows some of the effects of a muscle training program.

TABLE 12.1 — Beneficial effects of resistance training

Improved muscle strength and endurance
Increased bone mineral density
Improved ADL in older age
Improved insulin sensitivity
Possible improvements in resting blood pressure and percent body fat

Beginning a musculoskeletal fitness program during early to middle adulthood if one has not started earlier will help maintain fitness through the lifespan. Along with a cardiorespiratory fitness program, this helps delay the onset of age-related disabilities that limit quality of life. Improvements in muscular strength, endurance, and power all result from appropriate resistance training programs.

Resistance training methods vary in style from static isometric exercise to dynamic isotonic and isokinetic types of exercises in which concentric and eccentric movements are employed to provide the muscular resistance. When applied to resistance exercise a "rep" refers to one repetition of the complete start-to-end cycle of any particular resistance exercise (e.g., lifting and lowering a weight through the range of motion). A "set" refers to any series of reps performed one after the other without rest in between each rep. The force a subject can exert in completing one rep at maximal effort defines the one-rep-maximum, the "1 RM," for that specific exercise movement. "Intensity" for resistance training usually refers to the force exerted in completing one rep of an exercise as a percentage of the 1-RM. Given these definitions, there is considerable variability in the manner in which intensity, reps, and sets are combined in any particular resistance training program. In general, lighter resistances (low intensity) and more reps per set are prescribed to enhance muscle endurance contrasted with much heavier resistances (high intensity) and fewer repetitions to target strength development. For example, if endurance is the primary goal, resistance training can be applied in a circuit of several consecutive exercises using lighter resistances with relatively more repetitions targeting different muscle groups and allowing little rest between different exercise stations. In common to all resistance training methods is the principle of progressive overload, which states that an increase in strength (or endurance) requires that an overload or stress be applied to the muscle. As alluded to previously, this overload can be accomplished by increasing the intensity or the number of repetitions for each resistance training exercise, and follows the principle of specificity of training, in that a muscle or system adapts according to the stress imposed on it. As an example, performing low numbers of repetitions but using heavy resistance tends to build more strength, whereas increasing the number of repetitions at lighter weight tends to increase muscle endurance.

Given the wide range of training methods, it is not surprising that many different types of equipment have been devised for resistance training, ranging from free weights, to machines with weight plates or air-pressure resistance, to mechanical devices with motors and flywheels, to elastic bands and even plastic jugs, cans of soup or vegetables—anything that will provide a resistive force to the muscle to overload the contractile capability and initiate the adaptive response. A review of the merits and benefits of the various methods and techniques employed in resistance training programs is beyond the scope of this laboratory. Nor will sport-specific resistance training programs or periodization plans be a focus of this laboratory chapter. Instead, a process for devising a general resistance training program will be explored which could easily be included in a well-rounded exercise program to improve health and physical fitness for everyday living.

Before moving on to the procedures for prescribing resistance exercise to follow here, the student should review the general principles of exercise prescription presented in the laboratory chapter entitled "Exercise Prescription for Aerobic Fitness in Healthy Populations." The resistance exercise prescription for health-related fitness is an extension of the exercise prescriptive process. Both aerobic and musculoskeletal fitness training are often prescribed to be carried out concurrently and are integrated into the warm-up and conditioning phases of the planned exercise session

Of course, the emphasis on one or the other may vary from day to day in the training week, with resistance training on alternate days from aerobic training, or on the same day, depending on the time constraints and the goals/objectives of the subject. As an important side note, the adaptive response to a training program combining aerobic and resistance exercise is not well defined, and more research is needed to understand an apparent "interference effect" to assist the exercise practitioner in understanding the limitations of concurrent training. Current research suggests that resistance and aerobic training performed concurrently may result in training improvements of less magnitude, e.g., less VO_{2max} and strength improvements, compared with improvements measured when either training modality is performed alone, though individual responses vary widely (Glowacki et al., 2004; Karavirta, et al., 2011). This does not mean that health benefits are likewise affected, since no longitudinal research is available to address this issue. The general model for exercise prescription given previously in Chapter 11 can be modified as follows to accommodate resistance exercise in overall training program (Thompson, Gordon, and Pescatello, 2010).

1. **Mode**—the type or kind of exercise. For resistance training it is generally recommended that dynamic movements rather than static exercises be included in the exercise program, essentially any activities that overload the muscle throughout the joint range of motion. It is important that the selection of exercises target all the major muscle groups of the body: legs, hips, upper and lower back, shoulders, chest, and abdomen. Examples of multijoint exercises to include are leg press (squat), chest (bench) press, shoulder press, pull-ups (or pull-downs on machines), abdominal curl-ups, back extension exercises, and upright or bent rowing. Single-joint exercises are useful to isolate smaller muscle groups around a joint such as leg curls, quadriceps extensions, calf raises, biceps curls, triceps extensions. If resistance-exercise equipment is not readily available, calisthenics, which use the body weight as the resistive force and targetspecific muscle groups, can be successfully applied to improve musculoskeletal fitness. For practical benefit, consider carefully and include exercises that exert balanced training on opposing muscle groups to avoid a relative over-development of any one muscle group.

2. **Frequency**—the number of exercise sessions each day or week. Health benefits are realized by resistance training all the major muscle groups 2 to 3 $d \cdot wk^{-1}$, but not the same muscle group on consecutive days. Recovery from resistance exercise is necessary for adaptive benefits to accumulate. Adequate recovery is usually considered to be 48 hr. between training sessions. It is not uncommon to incorporate a 4 $d \cdot wk^{-1}$ resistance training frequency by splitting the workouts into shorter daily durations and training only one segment of the body each day allowing 48 hr recovery between training any one muscle group, e.g., performing only upper body exercises days 1 and 3, and only lower body exercises days 2 and 4 of the training week.

3. **Volume of Resistance Exercise**—a way to quantify the amount of resistance exercise. It is common to use a shorthand notation for resistance exercise prescriptions as; sets × reps × weight lifted in pounds or kilograms. Thus, 2 sets of 12 reps per set using 50 pounds of weight would be written: 2 × 12 × 50 lb. When applied to quantifiable resistive forces, e.g., pounds of weight lifted, *Volume* can be defined as the total amount of weight lifted in all exercises in a workout session. Volume can also be used as a measure of the amount of weight lifted in each set calculated as: [weight lifted in pounds (or kg) × repetitions per set]. The values can then be summed over the number of sets to obtain a volume measure for that specific exercise. If the reps and weights are identical for each set, then volume can be calculated as [weight lifted per rep × number of reps × number of sets]. For example, using the notation above, 2 × 12 × 50 lb = 1200 lb. As noted previously, intensity for resistance exercise is generally gauged as a percentage of the 1-RM and measured in pounds (or kilograms) of resistive force applied in each lifting rep. The resistive force applied or weight lifted settings vary with the type of resistance equipment used, and there is not direct transfer to from one type to another. For example, a chest press of 100 pounds using a weight-plate machine does not directly transfer to a bench press of 100 pounds using free weights. The intensity and reps per set are inversely related, and the objective of the training program must be considered in setting the these volume components (Table 12.2) In general, high-volume

TABLE 12.2 — General objectives of resistance exercise and suggested training program components

Objective	Frequency d · wk^{-1}	Sets	Reps	Intensity % 1-RM
STRENGTH, Mass	3	3	≤6	≥80%
STRENGTH, MASS, Endurance	3	3	8-12	60%-80%
Strength, Mass, ENDURANCE	3	3	≥15	≤60%
Rest between sets should generally be 90 sec to 3 min					

NOTE: The objectives in CAPITAL letters are those primarily improved by the program.

resistance training, that is, high repetition sets with lower intensity, improves local muscular endurance whereas low volume training, sets with higher intensity but fewer reps, improves strength.

4. **Duration or Time**—the length of time the exercise should be continued each session or accumulated over a week. This is dependent on the objective of the exercise program and the balance between aerobic and resistance training goals and objectives. For example, since loss of muscle mass is associated with aging, it is logical that the resistance exercise component of an older person's exercise program might occupy a greater proportion of the conditioning time than in younger individuals. It is unrealistic to expect that a sufficient amount of time can be devoted to both aerobic and resistance exercise training in a 20-minute period, the lower end of time recommended by ACSM for the conditioning phase of a typical training session (Thompson, Gordon, and Pescatello, 2010). Either the number of exercise days per week or the time per session will need to be increased to accommodate adequate aerobic and resistance training.

5. **Progression/Maintenance**—the rate at which the exercise dose is increased or maintained. The strength training progression plan is grounded in the practical application of the progressive overload principle. Simply put, this principle means that to stimulate an adaptation, the muscle must repeatedly face an overload challenge. But too much overload will result in negative gains and potential injury. For healthy individuals seeking resistance training for health benefits, the progressive overload should be managed conservatively. One simple way to manage the progressive overload is to prescribe to the subject, after a beginning accommodation phase of resistance exercise, an initial 8 rep training weight for the first set of 3 sets. When the subject has progressed to the point they can complete 12 reps with the same weight in all 3 sets, increase the weight such that the person can only achieve an 8 rep first set. Once the subject's goal for resistance training has been met, the gains can be maintained without increasing the resistance and with as little as one exercise session per week, as long as the resistance intensity is not diminished.

Equipment Needed

Resistance training equipment

Stopwatch or timing device

TECHNIQUES AND PROCEDURES

It is assumed in this laboratory that resistance exercise will be prescribed as a component of an exercise prescription for health, and that the subject is an apparently healthy adult. It is also assumed that the student will have previously completed Laboratory 11, "Exercise Prescription for Aerobic Fitness in Healthy Populations." It would be most efficient if Chapter 11 and this resistance exercise prescription laboratory were completed together and using the same subject, since much of the data needed for this laboratory is also required for aerobic exercise prescription, and references to tables and procedures from Chapter 11 will be made frequently below. The student will note many similarities between the preliminary procedures for these two laboratory experiences.

PREPARTICIPATION SCREENING

1. If informed consent has not been obtained on this subject as a part of completing Chapter 11, it should be done now before proceeding with any of the prescription procedures. Informed consent documents easily adaptable for the exercise prescription process can be found in the laboratory chapter entitled "Submaximal Exercise Testing for Estimating Aerobic Capacity." Once the informed consent has been signed and witnessed, proceed with the preparticipation screening.
2. Ask the subject to complete the Personal Information, Health Information, and Goals and Exercise Preference sections of the Exercise Prescription Subject Questionnaire. If this completed questionnaire is available for this subject from Chapter 11, then a new questionnaire need not be completed here.
3. Review the Health Information section of the Exercise Prescription Subject Questionnaire (refer to Table 11.2 in Chapter 11 in evaluating this information), then complete the Health Information and Classification section of the Resistance Exercise Prescription Worksheet.
4. **Decision Point 1**. Review the description of the three health categories described previously in Chapter 11 to assign the subject to the appropriate category: (1) **Low risk**, (2) **Moderate Risk**, or (3) **High Risk**.
5. **Decision Point 2**. Review the ACSM recommendations presented in Table 11.3 of Chapter 11. Indicate whether or not this subject should be required to complete a medical exam and diagnostic exercise test prior to participation in the exercise program. Also indicate if they are cleared for participation in a moderate or vigorous exercise program. (Assume that this subject wishes to begin a *moderate* exercise program.) If SLM-GXT data are not available, use the data in Table 11.5 of Chapter 11 to gain experience prescribing exercise.
    a. **YES**, medical clearance is recommended. An exercise prescription should not be given until the subject is properly evaluated by a physician and has completed a diagnostic exercise test.
    b. **NO**, medical clearance is not required. Proceed through the exercise prescription process.
6. Complete the Resting and Available SLM-GXT Data section of the Resistance Exercise Prescription Worksheet.

PREPRESCRIPTION ASSESSMENT

In order to prescribe a resistance training program, it is imperative that the subject have an assessment of his/her initial fitness level. Using the techniques described in Chapter 3, "Testing of Muscle Strength, Endurance and Flexibility," it is important to evaluate the initial musculoskeletal fitness level of the subject. Tests should be done on the major muscle groups of both the upper and lower body. For general fitness, it is not as important to assess muscle fitness using very sport-specific motions, so a general battery of tests can be used. The one-repetition maximum (1-RM) lift, as described previously, is the most accurate assessment of muscle strength on which a prescription for resistance training can be based.

There are, however, prediction equations for estimating 1-RM that can be applied using lighter weights. If lighter weights are used, it is important to remember that the more repetitions employed, the more the assessment will be of muscle endurance instead of muscle strength. This is problematic for athletes engaged in serious strength training, but less so for those interested in fitness and health. An example of a prediction equation that is often used is that of Brzycki (1993):

$$1RM = \text{weight lifted} / [1 + (0.033\ N)]$$
where N = number of repetitions to failure

DEVELOPING THE EXERCISE PRESCRIPTION

When the assessment is complete a prescription for training can be made. Table 12.3 shows some general guidelines for performing resistance training exercises for fitness and health. These help the individual to maximize the training effect and minimize the danger inherent in improper technique.

1. Complete the warm-up and cool-down section of the Resistance Exercise Prescription Worksheet.
   a. Warm-up should last about 5–10 minutes and be structured to gradually increase the metabolic rate to that required for the conditioning phase of the workout. For resistance exercise, be sure to include static and dynamic stretching, and light to moderate calisthenics as part of the warm-up. The warm-up might also include very lightweight resistance exercises, using the same muscle groups to be trained in that particular session.
   b. Also prescribe cool-down activities lasting about 5–15 minutes. These can be similar to warm-up activities, but designed to gradually decrease the metabolic rate for recovery. End with stretching and relaxation exercises.
2. Record the resistance exercises to be performed and the assessed pretraining 1-RM (Pre 1-RM) in the spaces provided in the Resistance Baseline Data and Weekly Exercise Prescription section of the Resistance Exercise Prescription Worksheet.
3. Calculate the volume of exercise desired for Week 1, and record the prescribed sets, reps, and resistance for each exercise on the worksheet. Remember that the resistance should be set relatively low in the first 1 to 3 weeks, depending on the progress of the subject, to allow him or her to become accustomed to the exercise routine, and minimize muscle soreness.
4. Closely observe the subject in session 1 of week 1 to assess her or his ability to complete the session. Change the resistance for each exercise as necessary, either raising or lowering it depending on the ability of the subject, and ease or difficulty with which they completed resistance exercise session 1.

TABLE 12.3 — Useful tips for performing resistance training

- Exercises should be performed for each major muscle group 2 to 3 times per week.
- There should be a rest period of 48 hours between exercises involving the same muscle groups.
- Exercise training sessions can be divided, for example, into upper body on one day and lower body the next.
- If exercising all muscle groups on the same day, upper- and lower-body exercises should be rotated.
- Large muscle group exercises should be performed before small muscle group exercises on the same day.
- Multijoint exercises should be performed before single joint exercises.
- A spotter should be available at all times.

5. Progression considerations: How much should the volume as reps and resistance (intensity) be increased and how often? This depends on many factors including the age of the individual, his or her initial conditioning level and training experience, and personal goals. Those in very poor physical condition or older in age may need longer in the adaptation phase but will then progress relatively rapidly. Others who are younger and in adequate physical condition may tolerate a more rapid progression through the program.

    a. As mentioned above, the first 1 to 3 weeks is typically an accommodation phase for resistance training, with the volume, intensity, and resistance at a level that is relatively easy for the subject.
    b. After the accommodation phase, a common method to assess progression in adults performing resistance exercise for general health is the following. Determine a weight the subject can lift 12 times, but cannot do more. This would be called the 12-RM, designating this is the maximal amount of weight that can be lifted 12 times. The subject should use this weight for all three sets, even though early in training the number of reps in the second and third sets will likely be fewer than 12. An increase in intensity should be considered when the subject can perform 3 sets of 12 reps, completing the last set without undue fatigue. A new 12-RM can be assessed and the sequence begins again. Some coaches and scientists have advocated the "10% rule" in which it is suggested not to increase intensity or volume more than approximately 10% at a time.
    c. It is common to assess a new 1-RM after 6 to 8 weeks of training. This provides some information about how well the resistance training program is working for the subject, and provides a guide for prescribing the resistance for each exercise.

6. Maintenance phase. When the subject's goals have been achieved, the exercise prescription enters the maintenance phase. In this phase, there is no need to continue to increase the volume of training but simply to maintain the volume at the level achieved. Research has shown that strength can be maintained by as little as one training session per week, as long as the intensity is maintained at the final level achieved during the training phase.

7. Special Considerations: Is there anything that you know about the subject that would cause you to vary the exercise prescription? Some people have specific strength deficits in muscle groups, or from one side of the body to the other. This may be from accidents or illnesses, or simply from doing relatively little resistance exercise in the past on any one body region. For example, it is quite common for women to have stronger lower body, but weaker upper body musculature. These considerations need to be evaluated on a person-by-person basis to individualize their training program.

REFERENCES

American College of Sports Medicine. "Position Stand: Physical Activity and Bone Health." *Med. Sci. Sports Exerc.* 36 (2004): 1985–96.

Brzycki, M. "Strength Testing: Predicting a One-Rep Max from a Reps-to-Fatigue." *J Phys Educ Recr and Dance.* 64 (1993): 88–90.

Ehrman, J. K. (Ed.), *ACSM's Resource Manual for Guidelines for Exercise Testing and Prescription*, 6th ed. Philadelphia: Lippincott Williams and Wilkins, 2010.

Glowacki, S. P., S. E. Martin, A. Maurer, W. Baek, J. S. Green, and S. F. Crouse. "Effects of Resistance, Endurance, and Concurrent Exercise on Training Outcomes in Men." *Med. Sci. Sports Exerc.* 36 (2004): 2119–27.

Karavirta, L., K. Hakkinen, A. Kauhanen, A. Arija-Blazquez, E. Sillanpaa, N. Rinkinen, and A. Hakkinen. "Individual Responses to Combined Endurance and Strength Training in Older Adults." *Med. Sci. Sports Exerc.* 43 (2011): 484–90.

Nelson, M. E., W. J. Rejeski, S. N. Blair, P. W. Duncan, J. O. Judge, A. C. King, C . A. Macera, and C. Castaneda-Sceppa. "Physical Activity and Public Health in Older Adults: Recommendation from the American College of Sports Medicine and the American Heart Association." *Med. Sci. Sports Exerc.* 39 (2007): 1435–45.

Nieman, D. *Exercise Testing and Prescription: A Health-Related Approach*, 7th ed. New York: McGraw-Hill, 2011.

Pollock, M. L., B. A. Franklin, G. J. Balady, B. L. Chaitman, J. L. Fleg, B. Fletcher, M. Limacher, I. L. Pina, R. A. Stein, M. Williams, and T. Bazzarre. "Resistance Exercise in Individuals with and without Cardiovascular Disease: Benefits, Rationale, Safety, and Prescription—An Advisory from the Committee on Exercise, Rehabilitation, and Prevention, Council on Clinical Cardiology, American Heart Association." *Circulation* 101 (2000): 828–33.

Ratamess, N. A., B. A. Alvar, T. E. Evetoch, T. J. Housh, W. Ben Kibler, W. J. Kraemer, and N. T. Triplett. "Progression Models in Resistance Training for Healthy Adults." *Med. Sci. Sports Exerc.* 41 (2009): 687–708.

Thompson, W. R., N. F. Gordon, and L. S. Pescatello (Eds.). *ACSM's Guidelines for Exercise Testing and Prescription*, 8th ed. Philadelphia: Lippincott Williams and Wilkins, 2010.

EXERCISE PRESCRIPTION SUBJECT QUESTIONNAIRE

(Modified from Thompson, Gordon, and Pescatello, ACSM's Guidelines for Exercise Testing and Prescription, Philadelphia: Lippincott Williams & Wilkins, 2010.)

Personal Information

Name _____ Date _____ Phone _____

Address _____

Sex (M or F) _____   Height _____ in   Weight _____ lb   Age _____ yr

Health Information: Please check ONLY those that apply to you

____ 1. My doctor has told me that I have heart problems, and I should only do the physical activities recommended by my doctor.

____ 2. I often feel pains in my chest, or have abnormal heartbeats when I work or exercise.

____ 3. In the past month I have had chest pains when I was not doing any physical work or exercise.

____ 4. I have times when I feel faint, dizzy, or even pass out, or I have lost my balance when I became dizzy.

____ 5. My doctor has told me that I have high blood pressure or high cholesterol.

____ 6. I have bone or joint problems (e.g., back, knee, or ankle) that keep me from exercising or might be made worse if I exercise.

____ 7. I do not regularly do physical activity hard enough to make me sweat.

____ 8. I regularly take medications prescribed by my doctor.

    a. Give purpose: _____

    b. Name/Dose: _____

____ 9. I smoke or use tobacco products regularly.

____ 10. Women: I am now or might be pregnant.

____ 11. There are other reasons why I think I should not exercise or be careful exercising.

    a. List reasons: _____

NOTE: If any items are checked, the exercise prescription should be postponed until further review by a qualified professional. Medical clearance may be necessary before the individual begins to exercise or takes part in an exercise test.

Personal Exercise Goals and Preferences

1. In this section, please explain your personal goals for your exercise program. What are the main things you hope to gain by participating in regular physical activities? These may be things like: improve my overall health, increase my endurance, lose weight, meet new people, gain muscle strength, strengthen my bones, train for a specific sport or activity, etc. _____
   _____
   _____
   _____
   _____
   _____

2. What are the types of exercises and physical activities you would most like to do? List as many as you wish. These may be things like lift weights, walk, bicycle, jog, engage in a sport, etc. _____
   _____
   _____
   _____
   _____
   _____

RESISTANCE EXERCISE PRESCRIPTION WORKSHEET

Preprescription Information

Name: _____ Age: _____ Sex: _____ Date: _____

Height: _____ in   Weight: _____ lb

Health Information and Classification

CVD Risk Factors _____
Symptoms _____
Activity Habits _____
Conditions of Concern _____
Medications _____
Risk Classification: __ Low
                     __ Moderate
                     __ High
Refer for Medical Clearance? ____ Yes
                                   ____ No
Cleared for: ____ Moderate Exercise
               ____ Vigorous Exercise

Resting and Available SLM-GXT Data

Resting Heart Rate _____
Resting Blood Pressure _____ / _____
Max Heart Rate _____
Max Blood Pressure _____ / _____
Max Rate Pressure Product _____
Max Treadmill Time _____ min _____ sec
Max RPE _____ VO_{2max} _____
Max METs _____ FAI _____
Limiting symptoms _____
GXT Interpretation _____
Fitness Category _____

Warm-up and Cool-down Activities

Purpose	Activity Time (min)	Mode or Activities Types
Warm-up	5–10	
Cool-down	5–15	

Resistance Baseline Data and Weekly Exercise Prescription

	Resistance Exercises							
Pre 1-RM								
Week 1								
Sets								
Reps								
Resist								
Week 2								
Sets								
Reps								
Resist								
Week 3								
Sets								
Reps								
Resist								
Week 4								
Sets								
Reps								
Resist								
Week 5								
Sets								
Reps								
Resist								
Week 6								
Sets								
Reps								
Resist								
Week 7 Reassess								
New 1-RM								
Sets								
Reps								
Resist								

CALIBRATION OF THE TREADMILL

APPENDIX A

INTRODUCTION

Many treadmills are calibrated at the factory and the calibration is difficult to change. Some still allow calibration, and this appendix describes one of the simpler systems of calibration. Even if your treadmill cannot be calibrated for speed and/or grade, these can be checked. So, this section also describes how to check the speed and grade of your treadmill to determine if it needs calibration.

Equipment Needed

- Meter stick
- Level
- Masking tape
- Pencil
- Carpenter's square
- Small screwdriver

TECHNIQUES AND PROCEDURES

TEST PROCEDURE

To calibrate the treadmill, both the speed and the grade must be known. The principle is the same for almost every treadmill made, but adjustment screws are not always in the same place, so they must be located individually. As mentioned above, some cannot be calibrated except by technicians, but this will show you whether you need to have a person perform this service. First, the treadmill speed will be determined. In order to do so, proceed as follows:

1. Measure the exact length of the belt. This task may require more than one person if the treadmill is a large one. The simplest way to do this is to put a mark on the treadmill, then advance the belt manually and proceed to mark 1-meter intervals until you return to the original mark. Use a pencil for this task since chalk or felt-pen marks are wider and you may have an error in length measurement due to either ignoring the width of the mark or measuring the mark twice. Your measurement resolution should be to 1 mm. Record the belt length on your worksheet.

2. Place a small piece of adhesive or masking tape on one edge of the treadmill belt. The piece should be large enough so you can see it while the treadmill is moving. The tape will allow you to record the number of revolutions your belt makes.
3. Turn the treadmill power on. Using the speed control, set the treadmill to read the desired speed. The speed at which you calibrate should be similar to that at which you will run the exercise test. If you are going to use more than one speed in your test, set the speed at the highest speed to be used. This is the nominal speed and should be recorded on your worksheet.
4. While the treadmill is traveling at the set speed, count the number of belt revolutions in approximately 1 minute (i.e., how many times the mark on the belt passes a fixed point per minute). It is more accurate to count the exact number of revolutions and the exact time than to count the approximate number of revolutions in exactly 1 minute. Enter the number of revolutions and the time on your worksheet.
5. Convert to revolutions per minute. For example, if the belt made 41 complete revolutions in 56 seconds:

$$56 \text{ seconds} = \frac{56 \text{ seconds}}{60 \text{ seconds/min}} = .933 \text{ min}$$

$$\frac{41 \text{ rev}}{0.933 \text{ min}} = 43.944 \text{ rev/min}$$

Enter the number you obtain on your worksheet.
6. Multiply the number of revolutions per minute (step 5) by the measured belt length (step 1). This will give you the speed of the treadmill in m/min.

Example: Belt length = 2.532 m/rev

43.944 rev/min × 2.532 m/rev = 111.266 m/min

Enter the number you calculate on your worksheet.
7. To convert m/min to mile/hr, divide the answer in step 6 by 26.822 (m/min)/(mile/hr)

Example: $$\frac{111.266 \text{ m/min}}{26.822 \frac{\text{m/min}}{\text{mile/hr}}} = 4.15 \text{ mile/hr}$$

Place your number in the appropriate space in your work sheet.
8. The value obtained in step 7 is the actual treadmill speed in mile/hr. If the reading on the speed scale of the treadmill control box does not match this value, adjust the meter to the proper reading. On some treadmills, the calibration adjustment is accessible through a small hole in the front or back panel of the meter.
9. Your treadmill speed is now correct at the calibrated speed. To ensure complete speed calibration (i.e., speed scale is linear), set the treadmill at an intermediate speed and repeat steps 4 through 7. Since you calibrated at a higher speed, this setting should be accurate as well, and you now have calibrated through the range of speeds you will use in the test. If the speed is not correct, check all the steps on the first calibration. It is easy to count one too many or too few meters on the first step and

errors in arithmetic are not uncommon. If all of the steps are correct and you still cannot calibrate accurately at two speeds, the scale is not linear, and you should have a technician examine the speed controller.

When you have completed the calibration for speed, all that remains is to check the grade calibration. In order to do this, complete the following steps.

1. Place a level on the bed of the treadmill and adjust the treadmill grade to make certain that the treadmill is level. If the elevation or "grade" meter does not read "O" when the treadmill bed is level, adjust it so that it does.
2. Elevate the treadmill so that the % grade meter is at the desired level. It is common to use 20%, which is both a typical maximal grade in a walking test and one that results in easy calculations.
3. Using the carpenter's square and the level, measure the exact incline of the treadmill as shown in the diagram below. The level needs to be places along the top (long) edge of the carpenter's square. The slope or grade of the treadmill is equal to the rise divided by the run, which is the same as the long side of the square divided by the measurement on the short side. Write these numbers in the appropriate space in your worksheet.

$$\text{Example:} \quad \text{Rise} = 5.0''$$
$$\text{Run} = 22.5''$$
$$\text{Slope} = 5.0/22.5 \times 100 = 22.22\%$$

FIGURE A.1 – Diagram of treadmill grade calibration. Note, the readings for rise and run are taken from the inside edge of the carpenter's square.

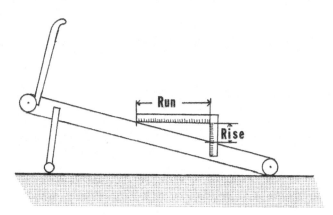

4. If you have the capability of doing so, use the small screw on the control panel for grade adjustment to adjust the screw until the % grade meter has the same reading as the one you obtained.
5. Verify your calibration by resetting the grade to 20% and repeating steps c and d. Then put the grade at 10% and check the calibration again. This process gives you a calibration at three grades (0, 10, and 20%), which should encompass the range of inclines that you will use in testing.

NOTE: Treadmill calibration checks should only take a few minutes. Therefore this is a procedure that should be performed at least once per week or whenever a round of testing is begun.

TREADMILL CALIBRATION WORKSHEET

Date: _____   Technician: _____

Speed Calibration

Belt Length: _____

	Trial 1	Trial 2	Trial 3	Trial 4
Nominal Speed (mile/hr):	_____	_____	_____	_____
Number of Revolutions:	_____	_____	_____	_____
Time (sec):	_____	_____	_____	_____
Rev/min: (number of rev/time/60)	_____	_____	_____	_____
Speed (m/min): (rev/min × belt length)	_____	_____	_____	_____
Actual Speed (mile/hr): [Speed (m/min)/26.822]	_____	_____	_____	_____

NOTE: If calibration is correct, nominal speed and actual speed should be identical.

Grade Calibration

	Trial 1	Trial 2	Trial 3	Trial 4
Nominal grade (%):	_____	_____	_____	_____
Rise (inches from rafter square):	_____	_____	_____	_____
Run (inches, from rafter square):	_____	_____	_____	_____
Actual grade (%): (rise/run × 100)	_____	_____	_____	_____

ALTERNATE PROCEDURE FOR TESTING TREADMILL GRADE CALIBRATION

___ 1. Place a carpenter's level on the treadmill bed and adjust the grade reading on the treadmill instrument panel to a setting of 0 (zero) percent.

    a. STOP: If the treadmill bed is not level, make adjustments to the treadmill legs according to the manufacturer's specifications until level is achieved. This is an essential step in evaluating the accuracy of the grade readings on the treadmill instrument panel. The treadmill bed must be level before proceeding to test the grade calibrations at other elevations.

___ 2. With the treadmill bed still set at zero percent grade, use a meter stick to measure two points from back to front length-wise on the top, outside edge of the treadmill deck (not on the belt) exactly 1-meter (1,000-mm) apart. This 1-meter measure can be anywhere along the length of the treadmill bed. It is usually most convenient to choose a length near the middle of the treadmill. Use small strips of masking tape to mark the front and back points of the 1-meter distance on the edge of the treadmill bed. Be sure to use an ink pen to mark on the masking tape the precise limits of 1-meter measurement. This will serve as a measure of the length of the **hypotenuse** of a right triangle used in the calibration process.

___ 3. Raise the treadmill so the instrument panel reads an elevation of five percent as a starting point to test the accuracy of the grade reading.

___ 4. Next, use the meter stick (or the long edge of the carpenter's level if it is marked in units of length), to measure the vertical height in millimeters from the floor to the front and back tape marks made on the treadmill edge in step 2 above. To ensure accuracy, it is very important that the meter stick (or carpenter's level) be *precisely perpendicular* to the floor surface in taking the vertical height measures. Holding the meter stick against the carpenter's level as the measure is taken can help ensure the meter stick is perpendicular to the floor.

___ 5. The **vertical rise** of the treadmill bed from back to front is the difference between these two vertical measurements. To calculate this, simply subtract the smaller from the larger measurement made in step 4.

___ 6. Calculate the percent grade can be calculated as follows. (Note that this procedure is accurate only for grades $\leq 20\%$.)

$$\text{Sine (angle of incline)} = \text{vertical rise} \div \text{hypotenuse}$$
$$(\text{sine} \approx \text{percent grade up to 20\%})$$
$$\text{If percent grade is} \leq 20\%, \text{then:}$$
$$\text{Percent Grade} = (\text{vertical rise} \div \text{hypotenuse}) * 100$$

NOTE: hypotenuse is constant 1,000 mm

___ 7. It is important to repeat the procedure at several percent grade settings. In addition to 0% and 5% suggested above, testing at additional grades of 10%, 15%, and 20% is advisable. Doing so will provide a calibration check at five grades, including zero grade (level). The grade settings evaluated for accuracy are chosen to span the range of inclines typically used in treadmill exercise testing.

___ 8. Record the measures in the Grade Calibration portion of the worksheet.

WORKSHEET FOR ALTERNATE TREADMILL GRADE CALIBRATION

___ 1. Treadmill is level when grade indicates "0" grade: Yes ____ No ____

___ 2. At 5% grade. Measure elevation from floor to upper mark treadmill bed: _____ inches

___ 3. At 5% grade. Measure elevation from floor to lower mark on treadmill bed: _____ inches

___ 4. Vertical rise = upper mark—lower mark

_____ = _____ cm − _____ cm

___ 5. Sine θ = vertical rise/hypotenuse

_____ = _____ cm/100 cm

Nominal Grade (%)	Upper Mark (cm)	Lower Mark (cm)	Rise (cm)	Sine θ
10				
15				
20				

CALIBRATION OF THE CYCLE ERGOMETER

APPENDIX B

INTRODUCTION

As with treadmills, calibration of many newer cycle ergometers is difficult and needs either a factory technician or expensive calibration equipment. Many nonelectric ergometers, though, can be calibrated. While some have a housing that covers the flywheel for safety, others do not, and some with a housing can still be calibrated. The technique described in this appendix is for calibrating Monark® cycle ergometers, which are widely used. The technique can be modified for most friction-braked cycle ergometers.

Ideally, the power absorbed by a machine should be measured while that machine is in use. This is called a dynamic calibration, but this sort of calibration necessitates the use of a dynamometer, which is large and expensive. So, we calibrate the ergometer statically, which is relatively simple, and much less expensive. The Monark cycle ergometer is mechanically braked and records resistance using a pendulum and scale. Some ergometers do not lend themselves to calibration, and thus should not be used in tests requiring accurate measurement of work. Electrically braked ergometers require specialized calibration equipment, whose demonstration is beyond the scope of this exercise.

Equipment Needed

Screwdriver
Weights (0.5 and 1.0 kg, total 7 kg)
Weight hanger

TEST PROCEDURE

On this type of ergometer, the force applied to the flywheel is equal to the difference between the forces pulling on the two ends of the friction belt. This difference is measured by the pendulum and scale on the ergometer. To calibrate this scale, follow the outlined procedures.

1. Detach the belt from the spring on the front of the ergometer. At this time it is a good idea to check the belt for signs of excessive wear (e.g., slick surface or frayed edges). If these are found, reverse or replace the belt if necessary.
2. Using the thumb screw in the front of the scale and with no tension on the resistance belt, align the mark on the pendulum with the "0" mark on the scale. When doing this be sure that the ergometer is on a level surface. If not, the pendulum will be vertical in relation to gravity, but not line up with the "0" mark.

3. Attach the weight holder to the spring. The actual design of the weight holder is not critical, but there are several critical elements to its design.

   a. It should be capable of supporting up to 7 kg of calibration weight.
   b. It should not rub against either the flywheel or any other part of the ergometer when it is suspended. This will result in misreading the weight.
   c. It should be short enough that when fully loaded, the bottom of the holder does not touch the ground.
   d. The weight of the holder must be known, since it is now part of the calibration weight. A simple way is to have the holder weigh exactly 0.5 kg.

4. Add weights to the holder. The weight of the holder plus the calibration weights should correspond to the number on the scale. Begin with an amount of weight that should reflect the highest you will use in the test. It is ideal to start with a large weight because any error in the scale is usually magnified at the high weights and thus easier to see and correct.

5. If the weights do not equal what is read on the scale, you will need to adjust the pendulum. On the end of the pendulum is a small weight secured in a cylindrical hole by a screw. When the screw is loosened the weight can slide freely in the hole. If the weight is slid closer to the end of the pendulum, it effectively makes the pendulum heavier at the end, so it requires more force to deflect the pendulum. If the weight is moved away from the end of the pendulum, the pendulum end is, effectively, lighter. If the weight on the scale reads more than the calibration weight, move the pendulum weight closer to the end of the pendulum until the scale accurately shows the weight of the holder plus the calibration weights, and secure the pendulum weight with the screw. Work in the opposite direction if your scale reads less than the calibration weight.

6. You have now calibrated the cycle ergometer scale at 0 kg and at a high force. Check your calibration using an intermediate weight as well to be sure that the scale is linear.

7. If your test requires that you use a specific resistance in order to perform an exact amount of work, it is best to calibrate the ergometer at that resistance and place a mark with tape on the resistance scale so that you can return to that exact resistance each time.

FIGURE B.1 – Diagram of 2 kg calibration weight suspended on Monark cycle ergometer and 3 kg reading of pendulum on scale.

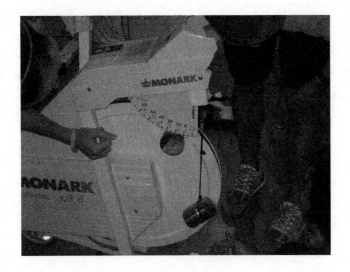

NOTE: Dirty bearings and chains can add resistance to pedaling that is not measured using the resistance scale. This could result in errors of which you would not be aware. It is therefore imperative that all bearings and chain links are kept clean and lubricated. A check of the chain and bearing resistance can be performed easily using a roll-down test. To do this, when the ergometer is known to be clean and well lubricated, detach the resistance strap and propel the pedals until the pedal rate is 60 RPM. Then remove your feet quickly from the pedals and record the time for the flywheel to stop turning. This is the criterion roll-down time for your ergometer. One new ergometer in this laboratory took over 15 min to come to a complete stop, but the time may vary with ergometer type. If the ergometer stops in less than about 75% of the original time, it is probably in need of lubrication and cleaning.

TYPICAL CALIBRATION OF OXYGEN AND CARBON DIOXIDE ANALYZERS

Most laboratories are equipped with automated metabolic carts that have manufacturer-specified procedures for calibration that vary widely with the specific product mode used but all generally require calibration of the volume monitor, O_2 and CO_2 analyzers. This calibration is imperative to ensure accurate results. Here we describe the calibration of a common set of stand-alone oxygen and carbon dioxide analyzers. In order to operate the analyzers, they should be turned on at least 20 minutes prior to use. After appropriate warm-up, the calibration of the analyzers can proceed. While the exact procedure for each individual analyzer may vary, most follow similar steps to those outlined below for the Ametek S-3A oxygen analyzer:

1. Check the cell temperature. It should read 6.790. If the number is not the same as this, adjust the temperature carefully with the TEMPERATURE ADJUST knob. This will take several minutes to adjust and it is very easy to "overshoot" your target, so be patient with the adjustment of temperature.
2. Turn the RANGE switch to $O_2\%$ and the REFERENCE ADJUST/UNKNOWN switch to REFERENCE ADJUST. Turn the REFERENCE ADJUST knob until the meter reads 20.9%, the value generally given for room air.
3. Calibrate the analyzer by running a gas of known O_2 concentration (usually 10%–15%) through the sensor. Place the REFERENCE ADJUST/UNKNOWN switch to the UNKNOWN position. Turn the CELL ZERO knob until the meter reads the correct $O_2\%$. The actual composition of the calibration gas is not very important, although it usually is one that approximates exhaled air. It is important, however, that the exact composition of the gas mixture is known.
4. Run a sample of the unknown gas through the sensor in the same manner as was used for the calibration gas. Read the $O_2\%$ from the meter.

The carbon dioxide analyzer, which uses infrared light to measure CO_2 concentration, is just as easily calibrated as the oxygen analyzer. The procedure for a typical CO_2 analyzer such as the Beckman LB-3 is as follows:

1. While sampling room air, adjust the ZERO knob until the output on the meter reads 0.02–0.05%, the approximate value for room air.
2. Introduce the calibration gas (5%–7% CO_2) into the sensor. Adjust with the GAIN knob until the meter reads the appropriate value.
3. Run a sample of an unknown gas through the sensor in the same manner as with the calibration gas and read the $CO_2\%$ from the meter.

If the analyzers do not calibrate correctly, there are several things that you might check. First make sure that during the calibration, the analyzer was sampling only the calibration gas and that there was no room air leaking into the sample. Be sure that the analyzers had time to warm up prior to beginning the calibration. If these do not improve the calibration, check with the operation manual for some further checks that can be made specific to the analyzer you use. Another concern with most gas analyzers is that they must be measuring gases without moisture in them. Moisture will both alter the composition of the gas because water vapor takes up some of the volume, and it will affect the analyzer, necessitating a thorough overhaul in some instances. Therefore, all gases that might have moisture in them have to pass through a desiccant prior to sampling.

METABOLIC CALCULATIONS FOR EXERCISE PHYSIOLOGY

APPENDIX D

This section is a compilation of useful equations and calculations for use in the exercise physiology laboratory. Some of these calculations are covered in chapters in the body of the laboratory manual, but the authors believe that having them in one place will be useful to the student.

PREDICTING MAXIMAL OXYGEN UPTAKE (ml/kg/min)
(Ref: Bruce et al., Am. Heart J., 85:546–562)

Active Men = 69.7 – 0.612 × (age in years)

Sedentary Men = 57.8 – 0.445 × (age in years)

Active Women = 42.9 – 0.312 × (age in years)

Sedentary Women = 42.3 – 0.356 × (age in years)

OR

Predicted $\dot{V}O_{2max}$ = 74.99 – 11.89 × (sex[+]) – 0.413 × (age in years) – 3.37 × (Physical Status[++])

Where: [+]Sex: Male = 1, Female = 2.

[++]Physical Status: Active = 1, Sedentary = 2.

FUNCTIONAL AEROBIC IMPAIRMENT

A. Definition: The <u>measured</u> $\dot{V}O_2$ of an individual expressed as a percentage of their age-predicted $\dot{V}O_2$; may be positive or negative.

B. Calculation:

$$FAI = \frac{[\text{Predicted } VO_2 - \text{Observed } VO_2]}{\text{Predicted } VO_2} \times 100$$

REGRESSION EQUATIONS FOR ESTIMATING $\dot{V}O_2$ MAX FROM TIME ON TREADMILL

BRUCE PROTOCOL (Foster et al., Cardiology, 70:85–93, 1983.)

$\dot{V}O_{2(ml/kg/min)}$ = 14.8 – (1.379 * $TIME_{min}$) + (0.451 * $TIME^2_{min}$) – (0.012 * $TIME^3_{min}$)

BALKE PROTOCOL

Men (Pollock et al., Am. Heart J., 92:39–46, 1976)

$$\dot{V}O_{2(ml/kg/min)} = (1.44 * TIME_{min}) + 14.99$$

Women (Pollock et al., Am. Heart J., 103:363–373, 1982)

$$\dot{V}O_{2(ml/kg/min)} = (1.38 * TIME_{min}) + 5.22$$

NAUGHTON PROTOCOL
(Foster et al., Cardiology, 70:85–93, 1983.)

$$\dot{V}O_{2(ml/kg/min)} = (1.61 * TIME_{min}) + 3.6$$

ACSM EQUATIONS FOR CALCULATING OXYGEN CONSUMPTION
(Ref. ACSM's Guidelines for Exercise Testing and Prescription, 8th ed., 2010.)

WALKING: Speeds 50 to 100 m/min (1.9 to 3.7 mph)

$$\dot{V}O_{2\,(ml/kg/min)} = (Speed_{m/min} * 0.1_{ml\,O2/kg/min/m/min}) + (1.8_{ml\,O2/kg/min/m/min} * Speed_{m/min} * fractional\ grade) + 3.5_{ml/kg/min}$$

RUNNING: speeds > 134 m/min (> 5 mph) and slower speeds of 80–134 m/min (3–5 mph) if truly running

$$\dot{V}O_{2\,(ml/kg/min)} = (Speed_{m/min} * 0.2_{mlO2/kg/min/m/min}) + (0.9_{mlO2/kg/min/m/min} * Speed_{m/min} * fractional\ grade) + 3.5_{ml/kg/min}$$

CYCLE & ARM ERGOMETRY

LEG CYCLE: power outputs between 50 & 200 Watts (300 and 1200 kp/m/min)

$$\dot{V}O_{2\,(ml/kg/min)} = (Work\ Rate_{kpm/min} * 1.8_{ml/kpm} \div Body\ Mass_{kg}) + 7_{ml/kg/min}$$

ARM CRANKING: power outputs between 25 & 125 Watts (150 & 750 kp/m/min)

$$\dot{V}O_{2\,(ml/kg/min)} = (Work\ Rate_{kpm/min} * 3_{ml/kpm} \div Body\ Mass_{kg}) + 3.5_{ml/kg/min}$$

Where:

$$Work\ Rate = Resistance_{kp} * Pedal\ Revolution\ Distance_{m/rev} * RPM_{rev/min}$$

Leg Ergometer Pedal Revolution Distance: **Monark = 6 m/rev, Tunturi = 3 m/rev**

STEPPING: for step rates between 2 and 30 steps/min

$$\dot{V}O_{2\,(ml/kg/min)} = (RATE_{steps/min} * 0.2_{ml/kg/min/steps/min}) + (HEIGHT_{m/step} * RATE_{steps/min} * 1.8_{ml/kg/min/m/min} * 1.33) + 3.5_{ml/kg/min}$$

EQUATIONS INVOLVING RELATIVE MEASURES OF O₂ CONSUMPTION

METs = $\dot{V}O_{2(mlO2/min)} \div body\ wt_{(kg)} \div 3.5_{(ml/kg/min/MET)}$ **OR** = $\dot{V}O_{2(mlO2/kg/min)} \div 3.5_{(ml/kg/min/MET)}$

Energy Expenditure$_{(kcal/min)}$ = **METs * Body wt**$_{(kg)}$ * **.0175**$_{(kcal/kg/min/MET)}$

Energy Expenditure$_{(kcal/min)}$ = $\dot{V}O_{2(LO2/min)} * 5_{kcal/L\,O2}$

RATE PRESSURE PRODUCT

A. Definition: This is an indication of the oxygen utilization of the myocardium.
B. Normal values at maximum exercise are: Men ≥ 325, Women ≥ 275
C. Calculation:

$$RPP_{max} = (SBP_{max} \times HR_{max}) \div 100$$

GENERAL EQUATION FOR SPIROMETRY CALCULATION OF $\dot{V}O_2$

$$\dot{V}O_{2(ml\ O2/min)} = (\dot{V}_{I(ml\ O2/min\ STPD)} * F_IO_2) - (\dot{V}_{E(ml\ O2/min\ STPD)} * F_EO_2)$$

DERIVATIONS OF THE FICK EQUATION

$$\dot{V}O_{2(ml\ O2/min)} = \dot{Q}_{(L\ blood/min)} * AVO_2\ diff_{(ml\ O2/L\ blood)}$$

$$\dot{V}O_{2(ml\ O2/min)} = HR_{(beats/min)} * SV_{(L\ blood/beat)} * AVO_2\ diff_{(ml\ O2/L\ blood)}$$

$$\dot{V}O_{2(ml\ O2/min)} = HR_{(beats/min)} * SV_{(L\ blood/beat)} * (CaO_{2(ml\ O2/L\ blood)} - CvO_{2(ml\ O2/L\ blood)})$$

$$\dot{V}O_{2(ml\ O2/min)} = HR_{(beats/min)} * SV_{(L\ blood/beat)} * [(1.34_{(ml\ O2/g\ Hb)} * [Hb]_{(g\ Hb/100\ ml\ blood)} * SaO_{2(\%\ O2\ sat.)} * 10)$$
$$- (1.34_{(ml\ O2/g\ Hb)} * [Hb]_{(g\ Hb/100\ ml\ blood)} * SvO_{2(\%\ O2\ sat.)} * 10)]$$

NOTE: multiplying by 10 in the calculation of blood O2 content converts ml O2/100 ml blood to ml O2/L blood. The unit associated with the "10" in the above equation is ml O_2/liter of blood/ml O_2/100 ml blood. The % O2 sat <u>must be decimal equivalent</u> of the percentage; e.g., 90% is 0.90 in the equation.

POPULAR PROTOCOLS

\multicolumn{5}{c}{BRUCE TEST}
Stage	Time (min)	Speed (mph)	m/min	Grade %
1	3	1.7	45.6	10.0
2	6	2.5	67.0	12.0
3	9	3.4	91.1	14.0
4	12	4.2	112.6	16.0
5	15	5.0	134.0	18.0
6	18	5.5	147.4	20.0
7	21	6.0	160.9	22.0
8	24	6.5	174.3	24.0
9	27	7.0	187.7	26.0
10	30	7.5	201.2	28.0

MODIFIED NAUGHTON TEST

Stage	Time (min)	Speed (mph)	m/min	Grade %
1	2	2.0	53.6	0.0
2	4	2.0	53.6	3.5
3	6	2.0	53.6	7.0
4	8	2.0	53.6	10.5
5	10	2.0	53.6	14.0
6	12	2.0	53.6	17.5
7	14	3.0	80.4	12.5
8	16	3.0	80.4	15.0
9	18	3.0	80.4	17.5
10	20	3.0	80.4	20.0

BALKE TEST: Constant 3.3 mph speed. Following a 1 minute warm-up at 0% grade, increase the grade to 2%, then 1% per minute thereafter until exhaustion.

Stage	Time (min)	Speed (mph)	m/min	Grade %
1	1	3.3	88.4	0
2	2	3.3	88.4	2
3	3	3.3	88.4	3
4	4	3.3	88.4	4
.	.	.	.	.
.	.	.	.	.
.	.	.	.	.

MODIFIED BALKE TEST: Constant 3.0 mph speed. After 3 min warm-up, increase grade 2.5% every 3 min.

Stage	Time (min)	Speed (mph)	m/min	Grade %
1	3	3.0	80.4	0.0
2	6	3.0	80.4	2.5
3	9	3.0	80.4	5.0
.	.	.	.	.
.	.	.	.	.

GENERAL LEG CYCLE ERGOMETER PROTOCOL

Warm up for 1 min at 50–60 rpm and workrate of 150 kpm/min (25 Watts). Increase 75 to 150 kpm/min (12.5 to 25 Watts) every 1–2 minutes thereafter until exhaustion.

DEFINITIONS OF SYMBOLS

$\dot{Q}$ = cardiac output

AVO$_2$ diff = difference in volume of oxygen between arterial and venous blood

$\dot{V}O_2$ = oxygen consumption

$\dot{V}O_{2max}$ = maximum oxygen consumption

HR = heart rate

SV = stroke volume

[Hb] = concentration of hemoglobin

CaO$_2$ = content (volume) of oxygen in arterial blood

$\dot{V}I$ = volume of inspired air per minute

$\dot{V}E$ = volume of expired air per minute

CvO$_2$ = content (volume) of oxygen in venous blood

SaO$_2$ = saturation (% expressed as a decimal) of arterial blood with oxygen

SvO$_2$ = saturation (% expressed as a decimal) of venous blood with oxygen

F$_E$O$_2$ = fractional concentration (% expressed as a decimal) of oxygen in expired air

F$_I$O$_2$ = fractional concentration (% expressed as a decimal) of oxygen in inspired air

CPSIA information can be obtained
at www.ICGtesting.com
Printed in the USA
LVOW02s0055180616
492814LV00004B/6/P

9 781465 219312